120 **Topics in Current Chemistry**

Fortschritte der Chemischen Forschung

Managing Editor: F. L. Boschke

Hydrogen Bonds

Guest Editor: P. Schuster

With Contributions by
A. Beyer, Th. R. Dyke, A. Karpfen,
C. Sandorfy, P. Schuster

With 33 Figures and 35 Tables

Springer-Verlag
Berlin Heidelberg GmbH
1984

This series presents critical reviews of the present position and future trends in modern chemical research. It is addressed to all research and industrial chemists who wish to keep abreast of advances in their subject.

As a rule, contributions are specially commissioned. The editors and publishers will, however, always be pleased to receive suggestions and supplementary information. Papers are accepted for "Topics in Current Chemistry" in English.

ISBN 978-3-662-16004-6   ISBN 978-3-540-38732-9 (eBook)
DOI 10.1007/978-3-540-38732-9

Library of Congress Cataloging in Publication Data. Main entry under title: Hydrogen bonds.
(Topics in current chemistry = Fortschritte der chemischen Forschung; 120)
Bibliography: p. Includes index.
Contents: Energy surfaces of hydrogen-bonded complexes in the vapor phase / A. Beyer, A. Karpfen, P. Schuster — Vibrational spectra of hydrogen bonded systems in the gas phase / C. Candorfy — Microwave and radiofrequency spectra of hydrogen bonded complexes in the vapor phase. Th. R. Dyke.
1. Hydrogen bonds — Addresses, essays, lectures. I. Schuster, P. (Peter), 1941–. II. Series: Topics in current chemistry; 120.
QD1.F58 vol. 120 [QD461] 540s [546'.2524] 83-14728

Managing Editor:

Dr. *Friedrich L. Boschke*
Springer-Verlag, Postfach 105280, D-6900 Heidelberg 1

Guest Editor of this volume:

Prof. Dr. *Peter Schuster*, Universität Wien,
Institut für Theoretische Chemie und Strahlenchemie,
Währingerstraße 17, A-1090 Wien

Table of Contents

Energy Surfaces of Hydrogen-Bonded Complexes in the Vapor Phase

Dedicated to Prof. D. Hadzi

Anton Beyer, Alfred Karpfen and Peter Schuster

Universität Wien, Institut für Theoretische Chemie und Strahlenchemie, Währingerstr. 17, A-1090 Wien, Austria[1]

Table of Contents

1 This work has been supported financially by the "Austrian Fonds zur Förderung der Wissenschaftlichen Forschung", Projects No. 3388, 3669.

Anton Beyer, Alfred Karpfen and Peter Schuster

1 Introduction

During the past few years a large number of review articles on hydrogen bonding were published in order to collect the enormous material which had been accumulated in numerous theoretical and spectroscopic works. In this article we concentrate on a few selected examples which are representative and which have been studied carefully enough to allow a straightforward comparison of theoretical and experimental data. A more complete listing of the results is, for example, given in references [1-5].

In section 2 we discuss some problems concerning the energy surfaces of intermolecular complexes. Due to their enormous flexibility hydrogen bonded aggregates present a challenge to both the theorist and the spectroscopist. Large amplitude motions have been observed in most hydrogen bonded vapour-phase complexes. The results obtained for some selected binary and ternary complexes as well as inifinite chains are discussed in sections 3 and 4. In the latter we try also to give an explanation for the non-additivity of intermolecular forces observed in hydrogen bonded systems. The final section summarizes general features of hydrogen bonds.

2 Energy Surfaces and their Accessibility by ab initio Calculations

Energy surfaces of molecules or molecular aggregates are central issues of theoretical chemistry and molecular spectroscopy. The Born-Oppenheimer approximation thus represents the basic concept in these fields since it provides the definition of an energy surface. As far as our present knowledge is complete the Born-Oppenheimer approximation is valid to a high degree of accuracy for molecules and molecular aggregates in the electronic ground states. Only few exceptions are known. In these cases the energy surfaces of the ground state and the first excited state come very close to each other in the neighbourhood of crossings of the "diabatic" energy surfaces. The energy surface of a molecular system in the electronic ground state $E_0(\mathbf{R})$ is defined by means of the stationary electronic Schrödinger equation

$$H\Psi_0 = E_0(\mathbf{R})\,\Psi_0 \tag{1}$$

where the vector $\mathbf{R}$ represents a complete set of internal coordinates. Such a complete set can be obtained from the set of all possible nuclear displacements by subtraction of the motion of the center of mass (CM) and the rotations of the molecular system. The Hamiltonian H is the operator for electronic motion in the potential of the entire molecule or molecular aggregate at fixed positions of the nuclei. For ν electrons and n nuclei we have

$$H = -\frac{1}{2}\sum_{\lambda=1}^{\nu}\vec{\nabla}_\lambda^2 + \sum_{\lambda=1}^{\nu-1}\sum_{\mu>\lambda}^{\nu}\frac{1}{r_{\lambda\mu}} - \sum_{\lambda=1}^{\nu}\sum_{i=1}^{n}\frac{Z_i}{r_{i\lambda}} + \sum_{i=1}^{n-1}\sum_{j>i}^{n}\frac{Z_iZ_j}{R_{ij}} \tag{2}$$

Electronic coordinates are denoted by Greek letters; r and R are the distances between the two particles named by the indices. Ψ_0 represents the electronic wave function, $\varrho(x, y, z)$ the electronic density distribution.

$$\varrho(x, y, z) = \int \dots \int \Psi_0^* \Psi_0 \, d\sigma_i \, d\tau_2 \dots d\tau_v \qquad (3)$$

τ_λ denotes the complete set of coordinates for electron "λ": $\tau_\lambda = x_\lambda y_\lambda z_\lambda \sigma_\lambda$ with σ_λ being the spin coordinate.

In general, a minimum of the energy surface corresponds to a set of stationary vibrational states of the molecular system. The position of the energy minimum is commonly called the equilibrium geometry $\mathbf{R_e}$. Analogously, we denote the expectation values for the molecular geometries in the vibrational states 0, 1, 2, ... by $\mathbf{R_0}$, $\mathbf{R_1}$, $\mathbf{R_2}$, etc. In most cases $\mathbf{R_0}$ is very close to $\mathbf{R_e}$. There are also exceptions to this correspondence which are important in the theory of intermolecular forces. We distinguish several cases:

1. Some potential wells are too shallow to sustain stationary vibrational states. The best known example of this kind is the dimer of helium He_2 (Fig. 1).

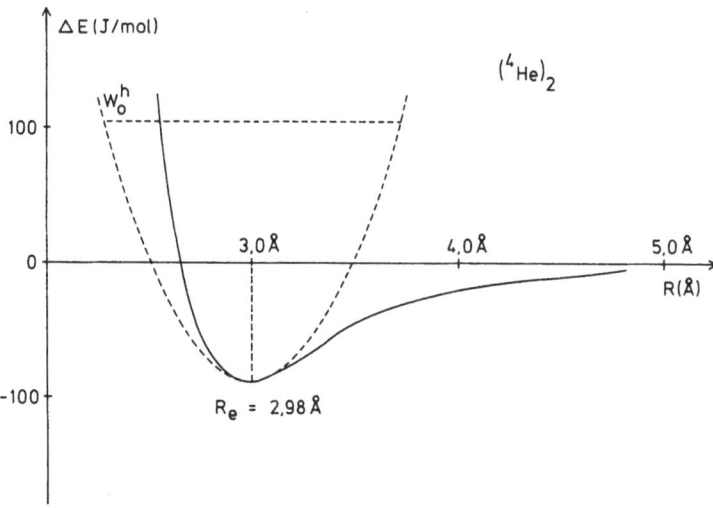

Fig. 1. Potential curve for ($^4He)_2$ (————). There exists no bound state for this complex. Harmonic potential curve (————) and corresponding vibrational ground state

The second case which is also characterized by large differences between equilibrium ($\mathbf{R_e}$) and zeropoint ($\mathbf{R_0}$) geometries occurs around energy minima the surrounding of which is very flat in a restricted region. Then, we find stable stationary states but large amplitude motions occur in one or eventually more internal degrees of freedom. Considering the well studied examples from intermolecular associations, we may divide the second case into two subclasses:

2. The large amplitude mode corresponds to a bending vibration or even a hindered internal rotation. The energy surface is then characterized by a more or less

3

extended valley with a rather flat bottom. Examples for large amplitude bending modes are the complexes $(HF)_2$, $H_2O \cdot HF$, $(H_2O)_2$ and many other binary associations of this type. We present the first two cases in fig. 2 because they are the best studied mobile hydrogen bonded complexes. In the case of hindered internal rotations the flat valley of the energy surface folds back onto itself, thus forming a closed orbit along which the energy is close to its minimum value. Examples for hindered internal rotations are the complexes formed by a noble gas atom and a hydrogen halide molecule (Fig. 3). The distinction between large amplitude bending vibrations and hindered internal rotations is not so superficial as it might appear at a first glance. In complexes of two polar molecules the leading term in an expansion of intermolecular energies is usually the electrostatic energy which becomes repulsive around certain orientations (Fig. 4). Internal rotation in these cases does not occur

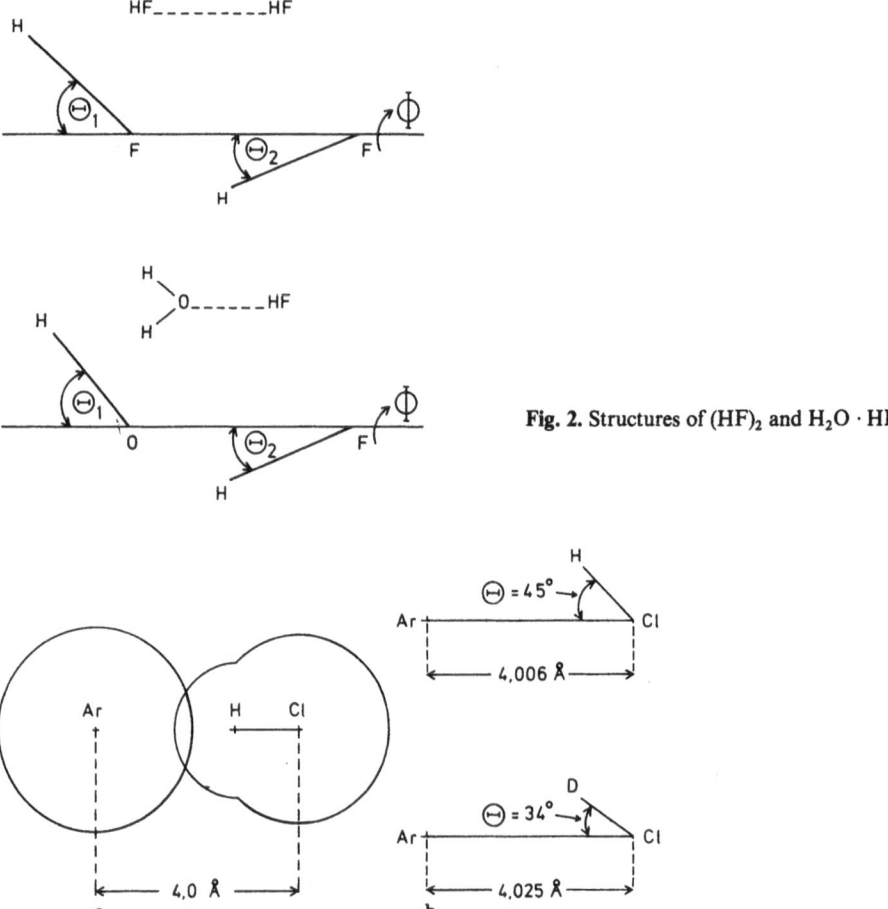

Fig. 2. Structures of $(HF)_2$ and $H_2O \cdot HF$

Fig. 3. a Equilibrium geometry; b ground-state geometry of Ar–HCl, the best studied complex of a noble gas atom and a hydrogen halide molecule. Interestingly the replacement of H by D leads to a substantial change in the geometry. Van der Waals' radii in a) are $R_H = 1.2$ Å, $R_{Ar} = 1.9$ Å

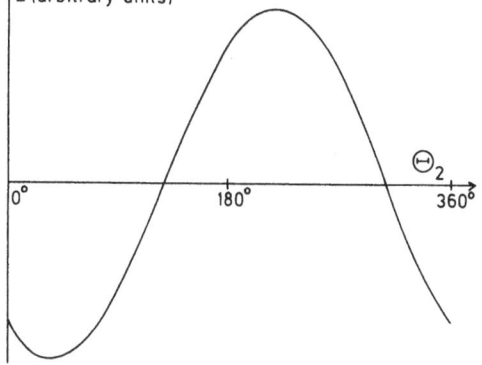

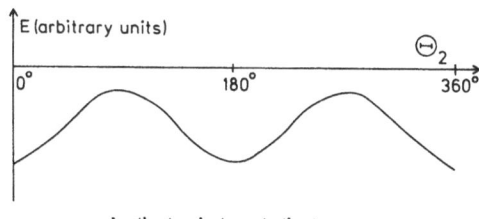

Fig. 4. a Interaction of two polar molecules

$$\Delta E = -\frac{\mu_a \mu_b}{R^3} \times (2 \cos \theta_a \cos \theta_b - \sin \theta_a \sin \theta_b \cos (\varphi_a - \varphi_b))$$

For the definition of θ_a and θ_b see Fig. 1. $\theta_A = \theta_1$, $\theta_b = (\theta_2 + 180°)$; b interaction between an atom and a polar molecule

$$\Delta E = \frac{\alpha_a \mu_b^2 (3 \cos^2 \theta + 1)}{2R^6}$$

μ_a, μ_b = dipole moments of molecules a or b, α_a = polarizability of atom a

since it leads to complex dissociation. In complexes of atoms and polar molecules, however, the leading terms are polarization and dispersion energies both of which are binding at all orientations. Consequently, a circular energy valley is sustained.

3. Occasionally, potential energy curves for bond stretching vibrations can be unusually flat as well. One special case is of particular importance in hydrogen-bonded complexes. At infinite intermolecular distance two states are possible in which the proton is either bound to molecule A or to molecule B. The two states are related by a proton transfer process:

$$AH + B \rightleftarrows A^{(-)} + HB^{(+)} \tag{4}$$

We start with the case where both reactants are electrically neutral molecules. The energy of the two oppositely charged ions on the r.h.s. of Eq. (4) is always larger than that of the two neutral molecules: reaction (4) is characterized by large positive reaction energies.

$$\Delta E_{PT} = E(A^-) + E(HB^+) - \{E(AH) + E(B)\} =$$
$$= E(HB^+) - E(B) - \{E(AH) - E(A^-)\} =$$
$$= \Delta E_{PA}(B) - \Delta E_P(A^-) \tag{5}$$

ΔE_{PA} denotes the proton affinities which are always negative quantities for neutral molecules and anions. Because of the negative charge, proton affinities of anions are larger in absolute value than those of neutral molecules and hence ΔE_{PT} is positive.

Anton Beyer, Alfred Karpfen and Peter Schuster

For proton transfer between small molecules the numerical values lie in the range $400 < \Delta E_{PT} < 1000$ kJ/mol (Table 1). The mutual approach of the molecular subsystems leads to complex formation. The two molecules on the l.h.s. of Eq. (4) form an ordinary hydrogen bond with the usual strength between $10 < \Delta E < 50$ kJ/mol. The complex on the r.h.s. of Eq. (4) is additionally stabilized by the Coulomb interaction of the two opposite charges. At the equilibrium geometry, or at the equilibrium geometries in case we have two minima of the energy surface, the position of the proton in the hydrogen bond depends on the balance between the proton transfer energy ΔE_P and the Coulomb-type stabilization of the ion pair. Depending on the position of the proton, whether the geometry of the isolated system AH or $HB^{(+)}$ is approximately preserved at the equilibrium geometry of the associate we distinguish neutral (NC) and ionic (IC) complexes,

$$A-H \ldots B \quad \text{and} \quad A^{(-)} \ldots H-B^{(+)} \tag{6}$$
$$\text{NC} \qquad\qquad \text{IC}$$

Table 1. Proton affinities and proton transfer energies in the vapour phase (kJ/mol)

$B + H^+ \rightleftarrows BH^+$			$A^- + H^+ \rightleftarrows AH$			$B + AH \rightleftarrows BH^+ + A^-$	
B	ΔH_{exp}	Ref.	A^-	ΔH_{exp}	Ref.	ΔH_{PT}	
He	−177.5	91)	J^-	−1290	99)	NH_3 + HJ	450
	−181.2	92)	Br^-	−1330	99)	NH_3 + HBr	490
Ne	−204.7	91, 92)	Cl^-	−1356	99)	NH_3 + HCl	516
Ar	−375.9	92)	F^-	−1532	99)	H_2O + HCl	643
Kr	−417.7	92)	OH^-	−1630	100)	H_3N + HF	690
Xe	−473.8	93)	H_2N^-	−1590	100)	NH_3 + NH_3	750
NH_3	−846 ± 8.5	94)	HS^-	−1410	100)	H_2O + H_2O	920
	−840	95)				H_2O + HF	825
PH_3	−784 ± 8.5	94)				HF + HF	1060
H_2O	−713 ± 8.5	94)					
	−707	95)					
H_2S	−728 ± 8.5	94)					
	−720	95)					
HF	−471.7	96)					
HCl	−569	97)					
HBr	−586	97)					
HJ	−615	98)					
CH_3OH	−762.6	94)					
C_2H_5OH	−781.4	94)					

A rough estimate of relative stabilities is shown in Table 1.

For most vapor-phase complexes of type (6) the Coulomb interaction compensates only part of the proton transfer energy ΔE_{PT} and we expect to find neutral type complexes (NC). The only candidates for ionic complexes (IC) or intermediate cases are the associations of strong bases with strong acids, e.g. the complexes between ammonia or aliphatic amines and HCl, HBr or HJ (Table 1).

Let us now consider the energy curve for the motion of the central proton at fixed

6

intermolecular distances (R_{AB}). At sufficiently large values of R_{AB} we find always double minimum potentials. The two minima correspond to the two types of associates, NC and IC. At some critical value of R_{AB} the double well collapses. The energy difference between the two minima, which represents the difference in stability between the neutral and the ionic complex before they collapse, determines the shape of the energy surface and hence also the dynamics of the association. If this energy difference is large, the higher energy minimum just disappears at the critical distance and the position of the lower energy minimum as well as the curvature of the energy surface around it are hardly affected. In case the energy difference is small we observe a substantial flattening of the potential curve for proton motion at the point of coalescence. The second energy minimum is appreciable dislocated. An extreme situation is encountered with symmetric hydrogen-bonded complexes where both minima have the same energy: at the point of coalescence both minima disappear and a new, very flat basin with the minimum in the middle of the two previous wells is found. Protons in such flat basins are extremely mobile and give rise to characteristic features in the vibrational spectra. So far, observations of these phenomena around the point of coalescence in symmetric or almost symmetric hydrogen bonds have only been reported for complexes in solution [6]. Crystal structure data on solids with symmetric hydrogen bonds give indirect hints on the existence of very flat energy minima [7]. The search for cases being intermediate between NC and IC in the vapor phase is of high actuality.

Almost all hydrogen-bonded complexes between neutral molecules in the vapor phase are of the neutral type (NC). The geometries of the subsystems in the equilibrium configuration of the complex are very close to those of the isolated molecules AH and B. The elongations of AH bonds involved in hydrogen bonding are extremely small, particularly in cases where A is a first row element (A = N, O or F, see Table 2). Ab initio calculations of low accuracy [8] seemed to give the general result that bond deformations are larger for proton donor molecules when A is a second or third row element (A = S, Cl, Br etc.). This finding correlates well with the stretching force constants of AH bonds in the corresponding isolated molecules. These are larger in the hydrides of first row elements (Table 2). Indeed, the force constant — and not the bond energy [2] — is the appropriate measure of the energy required for small deformations around equilibrium geometries. A larger force constant implies

Table 2. Elongation of HX bonds in hydrogen-bonded complexes

Y	X	near HF limit DZ + 2P			4-31G		
		R_{XY} (Å)	ΔR_{HX} (Å)	Ref.	R_{XY} (Å)	ΔR_{HX} (Å)	Ref.
HF	HF	2.830	0.005	[30]	2.69	0.005	[101]
H_2O	HF	2.718	0.011	[103]	2.61	0.017	[101]
H_2O	HCl	—	—		3.112	0.025	[101]
H_2O	HOH	3.027	0.004	[103]	2.832	0.007	[23]
H_3N	HF	—	—		2.779	0.027	[102]
H_3N	HCl	3.228	0.014	[16]			
HF	HOH	3.069	0.0	[103]			

that more energy is required to effect the same geometrical deformation. However, more elaborate calculations reveal (Table 2) that it is not completely sure whether such a systematic effect does exist. Because of the weak bond elongations more precise data are necessary to allow a definite conclusion to be made.

Due to the relatively small deformation in neutral complexes between uncharged molecules the frozen geometry approximation yields fairly correct results. In this simplified approach the geometries of the interacting subsystems are kept constant and only the internal degrees of freedom are optimized.

In the following we briefly discuss the quality of numerical calculations on the energy surfaces of intermolecular associations. Ab initio calculations for a given nuclear geometry (R) are attempts to find approximate solutions of Eq. (1). The quality of these calculations differs basically with respect to the structure of the wave function ψ_0 as well as the basis set within which the variational optimization is performed. We distinguish self-consistent field (SCF) calculations which use single determinant wave functions and configuration interaction (CI) calculations which include electron correlation effects by the use of many determinant functions for ψ_0. With increasing size (and quality) of the basis set, SCF calculations approach the single particle or Hartree Fock limit. CI calculations, in principle, allow an approach of the exact non-relativistic wave functions and energies. Technical difficulties, however, are prohibitive in most cases and one has to rely on defined truncations of the CI expansions. These truncations have characteristic names like independent electron pair approximation (IEPA) or coupled electron pair approximation (CEPA) and many others. For a review of the state of affairs in ab initio calculations of molecular systems see [9].

Accurate potential energy curves for dimers of noble gas atoms are available from precise scattering experiments and spectroscopic measurements [10]. Only in the case of He_2 do we have satisfactory agreement between theoretical and experimental data. So far, no calculated potential curve of Ne_2, which is comparable in accuracy with the experimental data, has been reported. The dispersion energy, the only attractive contribution to the two-atom potential, is not accessible at the Hartree Fock limit. Extensive and time-consuming CI calculations are indispensible in order to account properly for electron correlation effects. At the present stage of computational techniques complexes like Ne_2 seem to be just of a size which is still tractable with sufficient accuracy. For calculations on higher noble gas dimers the technical difficulties become enormous. Recently, a compromise between the non-empirical (ab initio) and an empirical approach has been proposed [11]: the repulsive contribution to the potential curve is taken from high-accuracy ab initio calculations at the Hartree-Fock level whereas dispersion energies are calculated from an empirical formula based on the R^{-n} expansion with empirical coefficients (n = 6, 8, 10). Energy curves thus obtained match well the most accurate experimental data.

Plenty of experimental information has been collected for the energy surfaces of atom-diatomic interactions. (see e.g. [5, 10, 12]) Most of the experimental studies were performed on systems like $Ar \cdot HCl$, $Ar \cdot HF$ or even heavier analogues for which the number of electrons is prohibitive for accurate ab initio calculations.

Concerning ab initio calculations on neutral hydrogen-bonded complexes we consider the simplest example, the hydrogen fluoride dimer (Table 3). Both the equilibrium geometry and energy of interaction strongly depend on the quality of the basis set

Table 3. Results of ab initio calculations on $(HF)_2$

Basis set (GTO's)	Method	R_{FF} (Å)	$R_{H_1F_1}$ (Å)	$R_{H_2F_2}$ (Å)	θ_1 (°)	θ_2 (°)	ΔE (kJ/mol)	Ref.
STO-4G	SCF	2.55	0.938[a]	0.938[a]	69	4	−21.76	61)
(9, 5, 1/4, 1) [4, 2, 1/2, 1]	SCF	2.80	0.917[a]	0.917[a]	52	5	−19.38	18)
(11, 7, 1/6, 1) [5, 4, 1/3, 1]	SCF	2.85	0.917[a]	0.917[a]	40	0[a]	−18.83	104)
(11, 7, 1/6, 1) [5, 4, 1/3, 1]	SCF	2.90	0.917[a]	0.917[a]	0[a]	0[a]	−18.33	104)
(11, 7, 2/6, 1) [7, 4, 2/4, 1]	SCF	2.90	0.902	0.917[a]	0[a]	0[a]	−14.48	105)
(11, 7, 2/6, 1) [7, 4, 2/4, 1]	CEPA	2.89	0.919	0.917[a]	0[a]	0[a]	−14.06	105)
(11, 7, 2/6, 1) [7, 4, 2/4, 1]	SCF	2.83	0.904	0.902	56.8	6	−15.90	30)
(11, 7, 2/6, 1) [7, 4, 2/4, 1]	CEPA	2.83[a]	0.904[a]	0.902[a]	56.8	6[a]	−17.16	30)
exp.		2.79			60–70		−25 ± 6.25	31)

[a] constrained values

Table 4. Results of ab initio calculations on $H_3N \ldots HCl$

Basis set GTO's	Method	R_{HCl} (Å) molecule	R_{NCl} (Å)	R_{HCl} (Å)	ΔE (kJ/mol)	Ref.
4-31G	SCF	1.274	3.13	1.274[a]	−45.20	106)
(11, 7/11, 7/6) → [5, 4/5, 3/3]	SCF	1.300	2.867	1.622	−79.42	107)
(12, 9, 1/9, 5, 1/5, 1) → [4, 3, 1/3, 2, 1/2, 1]	SCF	1.283	3.284	1.283	−33.0	108)
(12, 9, 1/9, 5, 1/5, 1) → [4, 3, 1/3, 2, 1/2, 1]	CI		3.228	1.323	−37.8	108)
(10, 6/8, 4/4) → [8, 4/5, 3/3]	SCF	1.305	2.857	1.719	−45.8	16)
(10, 6/8, 4/4) → [8, 4/5, 3/3]	SCF	1.305	3.006	1.386	−46.8	16)
(12, 8/10, 6/6) → [7, 5/6, 4/4]	SCF	1.294	2.880	1.755	−52.3	16)
(10, 6, 1/8, 4, 1/4, 1/4) [6, 4, 1/5, 3, 1/3, 1/3]	SCF	1.270	3.259	1.295	−35.85	16)
(10, 6, 1/8, 4, 1/4, 1/4) [6, 4, 1/5, 3, 1/3, 1/3]	CEPA	1.276	3.105	1.323	−45.1	16)
(12, 8, 2/10, 6, 2/6, 1) → [7, 5, 2/6, 4, 2/4, 1]	SCF	1.272	3.228	1.286	−25.2	16)

[a] constrained value

applied. Basis set truncation leads to essentially two main sources of errors in SCF calculations:

1. Basis set superposition in the complex results in an unphysical stabilization of the dimer. The isolated molecules are poorly described in case small basis sets are used. The quality of this description is improved when the basis set of the second molecule is added in the calculation of the dimer. This improvement results in a spurious stabilization of the aggregate. The basis set superposition effect has been studied in a number of calculations using "ghost molecules". These are basis sets of molecules without electrons and nuclei [13]. Such "counterpoise calculations" give upper limits for basis set superposition errors. Some authors used them as rough estimates of correction terms [14]. Indeed, it has been shown [15] that the spurious stabilization vanishes for large basis sets.

2. Some molecular properties which are of primary importance in calculations of intermolecular forces, like electric dipole moments, polarizabilities etc., are poorly reproduced by calculations with small basis sets. Evidently, these errors introduce errors into calculations of intermolecular forces, e.g. too large dipole moments yield too strong interactions. Table 3 shows convergence of the results on the $(HF)_2$ complex when a near Hartree-Fock quality of the calculation is reached.

Electron correlation introduces basically two effects into ab initio calculations on intermolecular forces. Hartree-Fock calculations do not account for dispersion forces and hence the dispersion energy is included only in CI calculations. A second contribution comes from a correction of monomer properties through electron correlation effects. Again, the correlation correction of the electric dipole moment is the most important contribution. In the case of $(HF)_2$ these two effects are of opposite sign and hence the influence of electron correlation on the calculated results is rather small (Table 3).

There are also complexes in which the position of the central proton is less obvious. The case of $H_3N \cdot HCl$, a possible candidate for proton transfer in the vapour phase, was found to be an especially tricky problem for ab initio calculations. Older theoretical studies using small basis sets favoured the ionic form (Table 4). More extensive calculations, in particular those with basis sets including several polarization functions on each atom, however, shifted the equilibrium position of the hydrogen atom strongly towards the chlorine atom. Compared to these drastic basis set effects on SCF calculations the influence of electron correlation on the calculated equilibrium geometry of the complex is rather small. The same is not true for the energy of interaction. The complex is substantially stabilized by electron correlation effects. From the best calculations available we conclude that the vapour-phase association of ammonia and hydrogen chloride comes close to the neutral complex (NC). The length of the hydrogen-chlorine bond in the complex is somewhat larger than that in the free molecule ($\Delta R_{HCl} \sim 0.05$ A). Complexes with mobile protons, thus, represent an enormous challenge for theoretical studies. Reliable results can be expected only from calculations with extended basis sets including sufficiently flexible polarization functions. Electron correlation is important as well, particularly for the energy of complex formation.

Still open is the question whether any complexes of the ionic type (IC) do exist in the vapour phase. Accurate ab initio calculations on the best candidates, $H_3N \cdot HBr$, $H_3N \cdot HJ$ and their alkyl derivatives like $(CH_3)_3N \cdot HBr$ etc., are not available yet. Studies in progress indicate strong elongation of the HBr bond in the complex $H_3N \cdot HBr$ [16]. There is an experimental indication that $(CH_3)_3N \cdot HBr$ is of ionic nature [17].

In the early seventies calculations of intermolecular forces became a fashion. Early progress in the field resulted in a real euphory as far as ab initio calculations on intermolecular complexes were concerned. Most of these studies were performed with small- and medium-size basis sets and gave approximate equilibrium geometries and energies of complex formation. Many of these calculations were repeated later on with larger basis sets and with electron correlation methods. Then it became clear that calculations accurate to a few tenth of a kcal/mol are extremely time-consuming. In the last few years calculations on intermolecular complexes became less numerous and the problems to be studied became more complex: the calculations did not stop at a few points around the equilibrium geometries and a value for the energies of inter-action. We mention some of these more extensive studies as representative examples: Yarkony et al. [18] calculated an extensive energy surface for hydrogen fluoride dimer in the frozen molecule approximation, Clementi and coworkers [19-21] performed an analogous and even more elaborate study on water dimer. Vibrational analysis on hydrogen-bonded complexes based on ab initio calculations has also become a routine for small- and medium-size basis sets: $(HF)_2$ [22], $(H_2O)_2$ [23]. Systematic studies on the ab initio calculation of thermodynamic functions were reported by Zahradnik and coworkers [24, 25]. Finally, a more accurate calculation of some vibrational constants of the complex $H_2O \cdot HF$ [26-28] including anharmonicity effects should be mentioned.

3 Binary Complexes of Hydrogen Fluoride and Water

Let us now turn to three examples of hydrogen-bonded complexes wich are known best at present: $(HF)_2$, $H_2O \cdot HF$ and $(H_2O)_2$. In Fig. 5 their equilibrium structures,

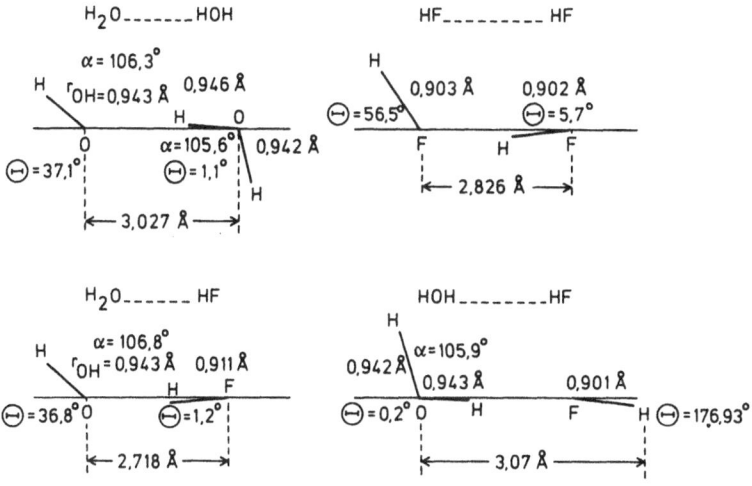

Fig. 5. Computed equilibrium structures of $(H_2O)_2$, $H_2O \cdot HF$, $(HF)_2$ and $HOH \cdot FH$

Table 5. Energies of interaction and thermodynamic properties of HF and H_2O dimers (energies in kcal/mol; reference state: 298.2 K, 1 atm)[a]

Proton acceptor	Proton donor	ΔE	ΔH_0^0	ΔH^0	$T\Delta S^0$	ΔG^0	K (atm^{-1})	References and comments
HF	HF	−4.1 / −5.5	−2.8	−3.2 / −4.3	−5.3 / −6.8	2.1 / 2.5	0.028 / 0.016	ΔE from [30]; rotational and vibrational contributions from [24]; gas-phase IR spectra and second virial coefficients from [67,68]
H_2O	HF	−7.5 / −7.2	−5.6 / −5.5	−6.3 / −6.2	−6.8 / −7.2	0.5 / 1.0	0.44 / 0.24	ΔE from [44]; rotational and vibrational contributions from [24]; gas-phase IR spectra from [51], T = 315 K
H_2O	H_2O	−4.8 / −5.4	−2.5	−2.9 / −3.6	−5.4 / −6.9	2.5 / 3.3	0.015 / 0.011	ΔE_{SCF} from [15], correlation energy from [21]; rotational and vibrational contributions from [24]; thermal conductivity studies of H_2O vapor [90], T = 373 K; data are more precise but in general agreement with those from second virial measurements

[a] Upper values; data from most extensive calculations; lower values: best experimental data available, thermodynamic functions calculated from partition functions by means of the Sackur-Tetrode equation

which were obtained by ab initio SCF calculations with identical basis sets, are compared.

The complex $H_2O \cdot HF$ is the most stable aggregate; the calculated energy of interaction is $\Delta E = -7.5$ kcal/mol. The two homodimers are of comparable stability: $(HF)_2$, $\Delta E = -3.8$ kcal/mol, and $(H_2O)_2$, $\Delta E = -4.0$ kcal/mol. The fourth combination $HF \cdot HOH$ represents the least stable structure: $\Delta E = -1.9$ kcal/mol. It has therefore not been observed so far in mixtures of HF and H_2O vapor.

Dissociation energies of vapor-phase dimers are of particular importance. Unfortunately, this quantity is neither easy to calculate nor accessible by direct experimental measurements. The difficulty with ab initio calculations lies in the enormous sensitivity of the energy difference to the choice of basis set and electron correlation effects (cf. [29]). Vibrational contributions are twofold: zero point energies and the temperature-dependent portions. The rotational partition function also contributes substantially to the temperature dependence [24]. In Table 5 are compiled the most accurate theoretical and experimental data on ΔE as well as on thermodynamic functions and the association constants. We observe general agreement. The most characteristic feature is the stability of the heterodimer $H_2O \cdot HF$. Its equilibrium constant is almost twenty times as large as those of the two homodimers. This fact makes direct studies on the heteroassociate in the vapor phase much easier than those on the two other complexes. Looking on the calculated and experimental values of ΔE more carefully, we find some systematic deviations for the two homodimers, for which the most accurate ab initio calculations have been performed so far: the calculated ΔE values are somewhat too small. This may well reflect a lack of optimization in the CI calculations of the complexes.

3.1 The Hydrogen Fluoride Dimer

Figure 6 compiles the theoretical and experimental data on the equilibrium geometry of $(HF)_2$. It reveals the commonly expected features of an ordinary hydrogen bond: the complex is planar and exhibits C_s-symmetry. A small elongation of the HF bond in the proton donor molecule relative to the bond in free HF is found. The inter-

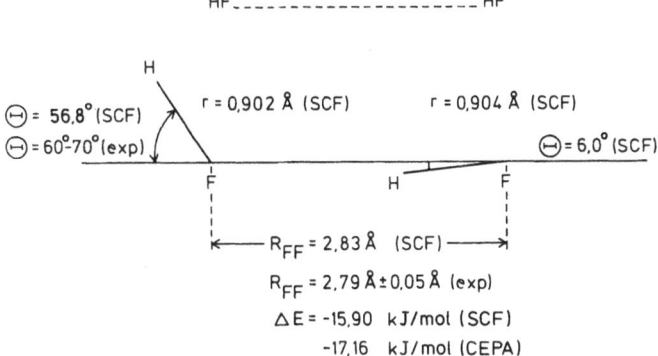

Fig. 6. Experimental and theoretical results on the structure of $(HF)_2$. The experimental results are taken from ref. [31] and the theoretical from ref. [30]

Table 6. Harmonic force constants of $(HF)_2$ (mdyn/Å)

Intramolecular force constants			Intermolecular force constants			
F_{XY}	4-31G [22]	(11, 7, 2/ 6, 1) [30]	F_{XY}	4-31G [22]	6-31G** [22]	(11, 7, 2/ 6, 1) [30]
F_{rr} (molecule)	9.55	11.17	F_{RR}	0.25	0.20	0.13
$F_{r_1 r_1}$	9.37	10.98	$F_{R\theta_2}/r_2$	0.017		−0.0076
$F_{r_2 r_2}$	9.17	10.76	$F_{R\theta_1}/r_1$	0.005		0.0074
			$F_{\theta_2\theta_2}/r_2^2$	0.18	0.15	0.12
			$F_{\theta_1\theta_2}/r_1 r_2$	0.054		−0.044
			$F_{\theta_1\theta_1}/r_1^2$	0.038	0.068	0.039
			$F_{\varphi\varphi}/R^2$	0.0004	0.0009	0.00012

molecular distance of $R_{FF} \sim 3$ Å is characteristic of ordinary hydrogen bonds between neutral molecules. In the other two complexes discussed here the hydrogen bond is not perfectly linear but there is a small but systematic deviation of $\theta_2 \sim 5°$ from the FF axis. The complex as a whole is strongly bent; the proton which is not involved in the hydrogen bond is displaced far from the FF axis. The axes of the two HF bonds intersect an angle of about $\theta_1 + \theta_2 = 60°$.

Harmonic force constants and vibrational spectra of $(HF)_2$ were calculated at the SCF level [22,30]. Numerical data are given in Table 6. They reflect the dynamical properties of this complex: the low values of the intermolecular force constants indicate high flexibility and large amplitude vibrations. Although $(HF)_2$ is planar at the equilibrium geometry, the rotation of one HF molecule around the FF axis changing the angle φ (Fig. 2) is almost free. A somewhat surprising result is the presence of a tunnelling motion which was first detected in the radiofrequency and microwave spectra of the complex [31]. Characteristic doublings of the rotational lines were observed. The energy surface of the complex gives indeed a strong indication of the nature of this tunnelling motion (Fig. 7). The cyclic structure of $(HF)_2$ with C_{2h} symmetry is a saddle point of the energy surface. The barrier of the tunnelling potential is roughly 1 kcal/mol. Two harmonic force constants are primarily involved in the potential of the tunnelling motion: $f_{\theta_1\theta_1}$ and $f_{\theta_2\theta_2}$, the former being only about one fourth of the latter. As expected, the proton-acceptor molecule is more "mobile" in the complex than the proton donor (for further details on the relative mobility of proton donor and proton acceptor see p. 19). It is interesting to note that the harmonic force constant of the change in inter-molecular distance f_{RR} has roughly the same value as that for hydrogen-bond bending ($f_{\theta_2\theta_2}$).

Both intramolecular force constants are lowered somewhat through complex formation (Table 6). As expected this effect is larger in the proton-donor than in the proton-acceptor molecule. In Table 7 we present calculated and experimental data on the vibrational spectrum of $(HF)_2$. General agreement is obtained. The most remarkable feature is the strict separation of intra- and intermolecular modes on the frequency axis. Hydrogen bond formation is a weak interaction compared to the formation of a chemical bond; hence, the normal frequencies are well separated. However, Hartree-Fock calculations of bond stretching force constants

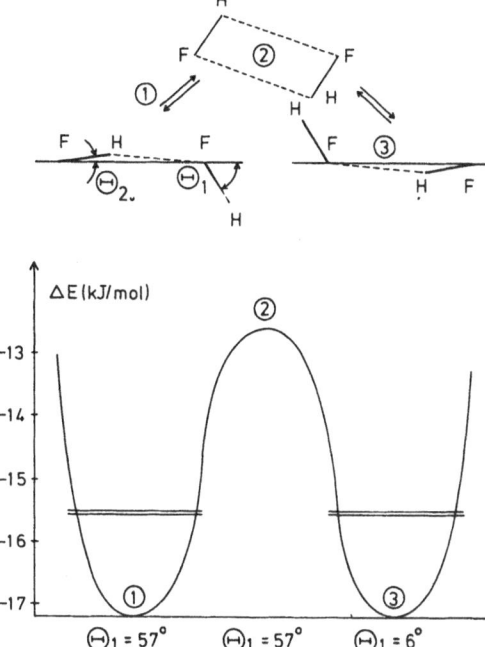

Fig. 7. Tunneling motion of the two protons in hydrogen fluoride dimer. The cyclic structure of the dimer corresponds to a saddle point on the energy surface

suffer from a general methodological deficiency. Due to the incorrect bond-dissociation behaviour of single determinant wave functions, these force constants are systematically too large which is reflected by too high frequencies of the corresponding normal vibrations. This error is seen in both the monomer and dimer spectrum. The frequency shifts are reproduced fairly correctly. It is possible to find small- or medium-size basis sets which reproduce experimental force constants quite well (see e.g. [32]). But these calculations yield the correct frequencies only by fortuitous compensation of errors and therefore are often quite bad with respect to other calculated quantities. The solution to the problem consists in very extensive large-scale CI calculations. These are extremely time-consuming and have not been performed yet.

There are some basic features of hydrogen-bonded complexes which are related to the wave functions and not to the energy surface of the complex. As an example we discuss in section 3.2 the increase in polarity on complex formation in the case of $(H_2O)_2$ and the experimentally well studied heterodimer $H_2O \cdot HF$.

3.2 The Water Dimer

$(H_2O)_2$ is nearly analogous to $(HF)_2$ although theoretical calculations and the evaluation of experimental data are much more difficult because there are two more atoms in the former complex. At the equilibrium geometry the dimer $(H_2O)_2$ has a plane of symmetry as well (C_s) (Fig. 8). The complex is strongly bent. The hydrogen atoms not involved in the almost linear hydrogen bond occupy trans positions with

15

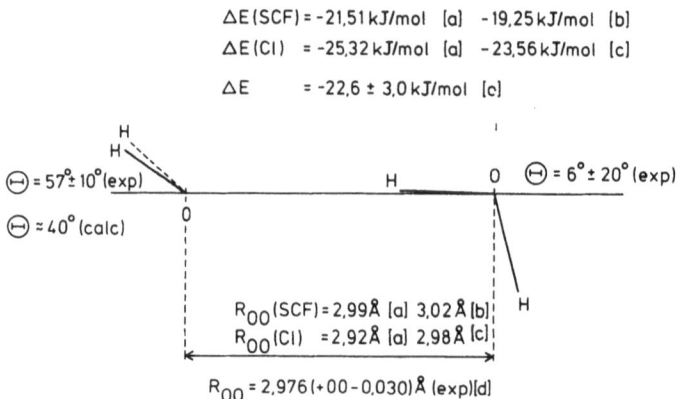

$$\Delta E\,(SCF) = -21{,}51\,kJ/mol \quad [a] \quad -19{,}25\,kJ/mol \quad [b]$$
$$\Delta E\,(CI) \;\; = -25{,}32\,kJ/mol \quad [a] \quad -23{,}56\,kJ/mol \quad [c]$$
$$\Delta E \qquad\quad = -22{,}6 \pm 3{,}0\,kJ/mol \quad [c]$$

$\Theta = 57^{\circ} \pm 10^{\circ}$ (exp)

$\Theta = 6^{\circ} \pm 20^{\circ}$ (exp)

$\Theta \approx 40^{\circ}$ (calc)

$R_{OO}(SCF) = 2{,}99\,\text{Å}$ [a] $3{,}02\,\text{Å}$ [b]
$R_{OO}(CI) = 2{,}92\,\text{Å}$ [a] $2{,}98\,\text{Å}$ [c]

$R_{OO} = 2{,}976\,(+00 - 0{,}030)\,\text{Å}$ (exp)[d]

Fig. 8. Experimental and theoretical results on the structure of $(H_2O)_2$; a) ref. [89], b) ref. [20], c) ref. [21], d) ref. [34] (for a discussion of basis-set superposition errors in these calculations see ref. [29])

respect to the $\overline{OO}$ axis. The intermolecular distance R_{OO} is about 3 Å. The equilibrium constant of dimerization of H_2O is too small to allow direct spectroscopic studies of the dimer in vapor phase, at least with the presently available techniques. However, sufficiently high dimer concentrations are obtained by the use of beam expansion in a supersonic nozzle. The structure of the complex has been determined by molecular beam electric resonance spectroscopy [33,34]. The experimental data agree well with the most accurate calculations available.

The $(H_2O)_2$ complex is highly flexible and performs large amplitude motions. As with $(HF)_2$ tunnelling occurs between two identical conformations. Because of the presence of two additional hydrogen atoms the path of the tunnelling motion is less obvious [35,36]. The first extensive calculation of the rotational and vibrational spectra of $(H_2O)_2$ as well as of the partition functions derived therefrom was performed by Braun and Leidecker [37]. Curtiss and Pople [23] presented an ab initio calculation of the harmonic force field of $(H_2O)_2$. The normal frequencies are compared with the available experimental data in Table 7. Unfortunately, no complete vibrational spectra have been reported for $(H_2O)_2$ in the vapor phase so far. There exists only fragmentary information on very low-frequency IR bands [38]. Several low-temperature matrix isolation studies have been reported [39-43]. The IR spectra depend on temperature and the matrix material. There seem to be several structures involved: the open-chain dimer (Fig. 8) and eventually also a cyclic dimer which represents the analogue of the transition state in the tunnelling motion of $(HF)_2$ in Fig. 7 [43]. Therefore, the assignment of experimental bands to calculated frequencies in [23] has to be considered with care. Nevertheless, there is no doubt that also in this case inter- and intramolecular frequencies are well separated. As far as the intramolecular force constants are concerned there is a small difference with respect to $(HF)_2$ since only the stretching force constants of the OH bond involved in hydrogen bonding is lowered. The other three OH bond stretching constants remain practically unchanged or are shifted towards slightly larger values [23].

The interpretation of the microwave spectra recorded in the molecular beam electric resonance experiments shows that the dipole moment increases on complex

Table 7. Vibrational frequencies of (HF)₂ and (H₂O)₂ (cm⁻¹)

(HF)₂	4-31G [22]	30)	exp. [109]
Stretching vibration of HF monomer	4117*	4452*	3962**
Stretching vibration of proton donor	4038 Δ = −79	4366 Δ = −86	3857 Δ = −105
Stretching vibration of proton acceptor	4081 Δ = −36	4418 Δ = −34	3895 Δ = −67
Intermolecular frequencies	171, 226, 588, 519	148, 185, 491, 436	
* harmonic frequency			
** harmonic frequency is 4139 cm⁻¹			

(H₂O)₂	exp. vapor [110]	exp. N₂ matrix [111]	SCF [23]	CI [36]
Monomer symmetric stretching	3657		3960	3832
antisymmetric stretching	3756		4098	3942
bending motion	1595		1767	1649
Dimer intramolecular stretching ν_3		3714 Δ = −11	4121 Δ = +21	3954 Δ = 12
ν_1		3626 Δ = −6	3979 Δ = +19	3840 Δ = 8
ν_b		3548 Δ = −84	3907 Δ = −53	3828 Δ = −4
ν_f		3698 Δ = −27	4085 Δ = −13	3941 Δ = −1
intramolecular bending ν_2'		1618 Δ = +21	1813 Δ = +46	1682 Δ = 33
ν_2		1600 Δ = +3	1771 Δ = +4	1664 Δ = 15
intermolecular			185, 204, 452	111, 125, 138
			81, 118, 536	167, 323, 583

Anton Beyer, Alfred Karpfen and Peter Schuster

formation. This increase in polarity is appropriately measured in terms of a dipole moment difference function:

$$\vec{\Delta\mu} = \vec{\mu}_{AB} - (\vec{\mu}_A + \vec{\mu}_B) = (\vec{\Delta\mu}_x, \vec{\Delta\mu}_y, \vec{\Delta\mu}_z) \qquad (7)$$

$$\vec{\Delta\mu} = (\vec{\mu}_{AB})_k - (\vec{\mu}_A)_k - (\vec{\mu}_B)_k; \qquad k = x, y, z$$

Thereby the monomers A and B are arranged in precisely the same geometrical positions which they occupy in the complex. The experimental data on the increase in dipole moments are in good agreement with the calculated data [44] (Table 8) provided the large amplitude motions are taken into account appropriately (see also [47]).

Table 8. Increase in electric dipole moments upon formation of hydrogen-bonded complexes (for the definition of $\vec{\Delta\mu}$ see Eq. (7))

Proton acceptor	Proton donor	$(\vec{\Delta\mu})_x$	$(\vec{\Delta\mu})_y$	$(\vec{\Delta\mu})_z$ [a]	Ref.
H_2O	HF	0.039	0	0.635	Best SCF calc. [44]
		—	—	0.68	Microwave spectrum [47]
H_2O	H_2O	−0.073	0	0.438	Best SCF calc. [44]
		—	—	0.45	Molecular beam electric resonance [34]

[a] All values are given in D. The z-axis coincides with the $\overline{OF}$ or $\overline{OO}$ connection line. The y-axis is chosen perpendicular to the plane of symmetry (xz)

3.3 The Water-Hydrogen Fluoride Complex

The complex $H_2O \cdot HF$ belongs to the best studied hydrogen bonded systems in the vapor phase. The fairly high association constant (Table 5) allows direct gas-phase studies to be performed in mixtures of hydrogen fluoride and water vapor.

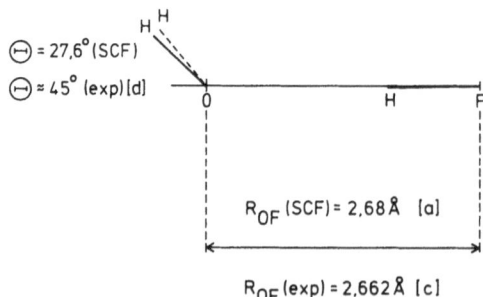

ΔE = 37,3 kJ/mol (SCF) [a]
ΔE = 30,0 kJ/mol (exp) [b]

⊖ = 27,6° (SCF)
⊖ ≈ 45° (exp) [d]

R_{OF} (SCF) = 2,68 Å [a]

R_{OF} (exp) = 2,662 Å [c]

Fig. 9. Experimental and theoretical results on the structure of $H_2O \cdot HF$; a) ref. [26-28], b) ref. [45]

The equilibrium geometry of the complex is shown in Fig. 9. It has C_s symmetry. As with the two homodimers $(HF)_2$ and $(H_2O)_2$ the complex as a whole is strongly bent: the C_2 symmetry axis of H_2O intersects the HF axis at an angle of $\theta_1 \sim 45°$. The hydrogen bond in $H_2O \cdot HF$ is somewhat stronger than in the two homodimers. Consequently, the intermolecular equilibrium distance is smaller: $R_{OF} = 2.68$ Å compared to $R_{FF} = 2.90$ Å and $R_{OO} = 3.00$ Å. The deviation from a perfectly linear hydrogen bond is less significant. The elongation of the HF bond on complex formation is still small ($\Delta R_{HF}(H_2O \cdot HF) = 0.011$ Å) but significantly larger than that in the proton donor in $(HF)_2$ ($\Delta R_H(HF \cdot HF) = 0.004$ Å) or the elongation of the HO bond in the hydrogen bond of water dimer ($\Delta R_{HO}(H_2O \cdot HOH) = 0.003$ Å). The experimental information on $H_2O \cdot HF$ has been provided by the marvelous microwave work of Millen and coworkers [45-47].

The heterodimer $H_2O \cdot HF$ performs large amplitude vibrations; the motion changing basically the angle θ_1 has been studied extensively in the microwave spectrum [45]. A symmetric double minimum potential with a barrier height of about 0.4 kcal/mol was obtained as the best fit to the lowest vibrational levels. There is no tunnelling motion in $H_2O \cdot HF$ since the other hydrogen-bonded complex, $HOH \cdot FH$, has a much higher energy. Experimental evidence of another large amplitude motion has been provided recently [48]: the mean amplitude of the hydrogen bond-bending motion was derived from nuclear quadrupole and nuclear spin — nuclear spin coupling in the rotational spectra of $H_2O \cdot DF$ and $H_2O \cdot HF$.

By this technique an average value $\langle \cos \theta_2^2 \rangle$ is measured. From this the authors calculated a mean amplitude $\theta_2 = 18.3°$ in $H_2O \cdot HF$ which is lower than the corresponding values for rare gas atom HF complexes: $Ar \cdot HF$, $\theta_2 = 41.1°$ and $Kr \cdot HF$, $\theta_2 = 39.2°$ [49,50]. As expected, hydrogen bonding introduces some rigidity into the complex: However, the isotope effect on θ_2 in $H_2O \cdot HF$ is not yet clear [48].

Vibrational frequencies of $H_2O \cdot HF$ were calculated by means of a quasilinear model which neglects angular modes. Nuclear motions are restricted to variations

Table 9. Vibrational frequencies of $H_2O \cdot HF$ (cm^{-1})

	SCF [28]	SCF [27]	CI [27]	exp [51]
Vibrational mode	$(10, 5, 1/5, 1) \rightarrow [5, 2, 1/2, 1]$	6-31G*	6-31G*	
v_{FH}	4206	4170	3927	3608
$v_{FH \cdots O}$	223	226	245	180 ± 30

in R_{OF} and R_{HF} [26-28]. Within this model the authors calculated the energy surface with an extended basis set including electron correlation effects. They derived a quartic force field including all 12 force constants. In Table 9 their results are compared with the gas-phase IR data recorded by Thomas [51]. Despite the neglect of angular motions the agreement is good. In the case of the closely related complex $(CH_3)_2O \cdot HF$ it was possible to obtain the first overtone in the IR spectrum as well [52] which gives important information on the details of the fine structure of the vibration-rotation bands in the fundamental region (see also ref. [119]).

In their detailed analysis of the microwave spectrum of $H_2O \cdot HF$ Kisiel et al. [45] derived a value for the increase in the dipole moment through hydrogen bonding (Table 8), taking into account the averaging effect of the large amplitude motion corresponding to θ_1. In contrast to previous results derived from the rigid planar structure, which gave much smaller experimental $\Delta\mu$ values (see Eq. (7)), there is an excellent agreement between calculated [44] and measured quantities.

4 Ternary Complexes, Clusters and Infinite Chains

A profound knowledge of the properties of higher aggregates, i.e. associations of more than two molecules, is of particular importance for a better understanding of interactions between molecules in dense vapors and in the condensed phases. In this section three-body forces and higher intermolecular interactions and subsequently the available data on higher aggregates are discussed. Infinite chains of hydrogen-bonded molecules will be briefly mentioned because these systems allow the most straightforward extrapolation from small clusters to condensed phases. Moreover, they are also proper reference states for the comparison of theory and experiment: infinite chains can be investigated by accurate abinitio calculations and they are good approximations to hydrogen halide crystals which consist of HF, HCl and HBr on p. 37).

In order to obtain a better understanding of the forces acting in molecular clusters we consider an appropriate partitioning of intermolecular energies. For the formation of an aggregate of these subsystems A, B and C

$$A + B + C \rightleftarrows ABC \qquad (8)$$

we may write

$$\Delta E = E_{ABC} - (E_A^0 + E^0 + E_C^0) = \Delta E_{AB} + \Delta E_{BC} + \Delta E_{CA} + \Delta E_{ABC} \quad (9)$$

E_{ABC} is the total energy of the ternary complex according to Eq. (1). It is, of course, a function of all internal degrees of freedom. E_A^0, E_B^0 and E_C^0 are the total energies of the three molecules A, B and C in the appropriate reference states, commonly the ground states. Binary interactions or pair potentials ΔE_{AB}, ΔE_{BC} and ΔE_{CA} have been discussed in the previous section. They are here the energies of interactions in the absence of the third molecule. ΔE_{ABC} is the contribution of the three-body forces. In higher aggregates four- five- and, generally n-body forces can be defined analogously. We shall now study the role of three- and higher n-body forces in hydrogen-bonded clusters. First, a world of pairwise interactions is visualized [5].

4.1 Additive Binary Potentials and the Role of Many-Body Forces

In order to derive positive indications of the importance of many-body contributions to structures of clusters we compare some very general results with experimental data from systems which have structures sufficiently simple to allow a straightforward interpretation. Some crystals will be well suited for this purpose.

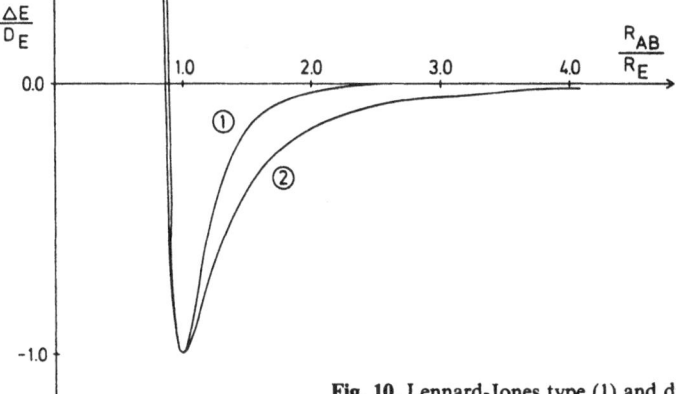

Fig. 10. Lennard-Jones type (1) and dipole-dipole (2) potential

In non-polar systems pair potentials can be simulated quite well by Lennard-Jones type functions:

$$\Delta E_{AB} = D_e \left\{ \left(\frac{R_e}{R_{AB}} \right)^{12} - 2 \left(\frac{R_e}{R_{AB}} \right)^6 \right\} \tag{8}$$

D_e is the depth of the well in the potential curve and R_e the equilibrium distance (Fig. 10). In the absence of many-body forces the energy of interaction in clusters is simply a superposition of expressions of type (8). For the trimer ABC, we have

$$\Delta E = D_e \left\{ \left(\frac{R_e}{R_{AB}} \right)^{12} + \left(\frac{R_e}{R_{BC}} \right)^{12} + \left(\frac{R_e}{R_{CA}} \right)^{12} \right.$$
$$\left. - 2 \left[\left(\frac{R_e}{R_{AB}} \right)^6 + \left(\frac{R_e}{R_{BC}} \right)^6 + \left(\frac{R_e}{R_{CA}} \right)^6 \right] \right\} . \tag{9}$$

It is straightforward to show that the equilibrium structure of the trimer is an equilateral triangle with $R_{AB} = R_{BC} = R_{CA} = R_e$ and $\Delta E = 3D_e$. Similarly, one finds for the tetramer ABCD that the most stable conformation is a tetrahedron with $R_{AB} = R_{BC} = R_{CD} = R_{DA} = R_{AC} = R_{BD} = R_e$ and $E = 6D_e$. In higher aggregates the situation becomes more complicated because the distances between individual pairs of molecules in the cluster can be no longer all the same. We have to distinguish between first, second, third etc. nearest neighbour interactions.

In order to illustrate the importance of these more distant interactions we choose a simple model system, a linear chain of molecules. We start with the trimer. From Eqs. (8) and (9) we obtain:

$$\Delta E^{(3)}(R) = D_e \left\{ 2 \left[\left(\frac{R_e}{R} \right)^{12} - 2 \left(\frac{R_e}{R} \right)^6 \right] + \left(\frac{R_e}{2R} \right)^{12} - 2 \left(\frac{R_e}{2R} \right)^6 \right\} \tag{10}$$

Thereby we have assumed equal nearest neighbour distances $R = R_{AB} = R_{BC} = R_{CA}/2$. R_e and D_e denote the equilibrium distance and dissociation energy of the dimer, respectively. Now, we search for the equilibrium geometry of the trimer

21

(under the contraint of linearity) by energy minimization $d\,\Delta E^{(3)}/dR = 0, R = R_e^{(3)}$. For the purpose of comparison we introduce a contraction factor f_3 which measures the difference in the equilibrium distance between dimer and trimer

$$R_e^{(3)} = R_e \cdot f_3; \qquad f_3 = \left(\frac{2 + 2^{-12}}{2 + 2^{-6}}\right)^{1/6}. \tag{11}$$

Clearly, we have $f_3 < 1$. The interaction with further neighbours is atrractive and leads to a contraction of intermolecular distances. The energy of the trimer at the equilibrium distance is easily calculated. For the purpose of comparison we introduce a dimensionless energy enhancement factor d_3 which refers to the energy per nearest neighbour interaction:

$$d_3 = \frac{\Delta E^{(3)}(R_e^{(3)})}{2D_e} = f_3^{-12}\left(1 + \frac{1}{2}\,2^{-12}\right) - 2f_3^{-6}\left(1 + \frac{1}{2}\,2^{-6}\right) \tag{12}$$

From Eqs. (11) and (12) we derive $d_3 < -1$ and hence we have increased stabilization through further neighbour interactions.

The same procedure can be applied to higher linear clusters of, let us say, n molecules. In order to retain the simplicity of the approach we have to assume equal nearest neighbour distances: $R_{AB} = R_{BC} = R_{CD} = \dots$. This is an approximation for clusters with $n > 3$, actually a very good one for Lennard-Jones systems. This assumption is not only exact for $n = 3$ but also for the infinite chain, $n = \infty$. By straightforward calculation we derive

$$R_e^{(n)} = R_e \cdot f_n;$$
$$f_n = \left(\frac{n - 1 + (n - 2)\,2^{-12} + (n - 3)\,3^{-12} + \dots + (n - 1)^{-12}}{n - 1 + (n - 2)\,2^{-6} + (n - 3)\,3^{-6} + \dots + (n - 1)^{-6}}\right)^{1/6} \tag{13}$$

and

$$d_n = \frac{\Delta E^{(n)}(R_e^{(n)})}{(n - 1)\,D_e} =$$
$$= f_n^{-12}\left(1 + \frac{(n - 2)\cdot 2^{-12}}{n - 1} + \frac{(n - 3)\cdot 3^{-12}}{n - 1} + \dots + \frac{(n - 1)^{-12}}{n - 1}\right) -$$
$$- 2f_n^{-6}\left(1 + \frac{(n - 2)\cdot 2^{-6}}{n - 1} + \frac{(n - 3)\cdot 3^{-6}}{n - 1} + \dots + \frac{(n - 1)^{-6}}{n - 1}\right) \tag{14}$$

Now, we perform the limit $n \to \infty$ and obtain a measure of the sum of all further neighbour effects in an infinite linear chain of Lennard-Jones systems:

$$R_e^{(\infty)} = R_e \cdot f_\infty; \qquad f_\infty = \left(\frac{\zeta(12)}{\zeta(6)}\right)^{1/6} \tag{15}$$

and

$$d_\infty = \lim_{n \to \infty} \frac{\Delta E^{(n)}(R_e^{(n)})}{(n-1) D_e} = -\frac{\zeta(6)^2}{\zeta(12)} \tag{16}$$

Herein, we made use of the Riemann-Zeta-function:

$$\zeta(k) = \sum_{n=1}^{\infty} n^{-k} \tag{17}$$

Some numerical results are given in Table 10. We observe rather small effects: the contraction of the equilibrium distance is 0.3% only and the energy per nearest neighbour interaction increases in absolute value by 3.5%. In three dimensions the effects are somewhat larger. The expressions for f_n, d_n, f_∞ and d_∞ are free of parameters and valid for all Lennard-Jones systems, independently of the particular values of R_e and D_e.

Since we are basically interested in polar systems here we make the same estimate with a simple binary potential function which accounts for polar interactions. A Stockmayer potential, which describes the interaction between two dipoles, in addition to a Lennard-Jones potential is most suitable for this purpose

$$\Delta E_{AB} = C_1 \cdot R_{AB}^{-12} - C_2 \cdot R_{AB}^{-6} - \mu_A \mu_B \cdot R_{AB}^{-3} \cdot g(\theta_A, \theta_B, \varphi) \tag{18}$$

C_1 and C_2 are constants, μ_A and μ_B represent the electric dipole moments of molecule A and molecule B, respectively. The angular dependence is summarized in the function

$$g(\theta_A, \theta_B, \varphi) = 2 \cos \theta_A \cos \theta_B - \sin \theta_A \sin \theta_B \cos \varphi \tag{19}$$

In order to facilitate our considerations we assume planarity, i.e. $\varphi = 0$. We restrict ourselves to symmetric arrangements with stabilizing dipole-dipole interactions: $\theta_A = \alpha$, $\theta_B = -\alpha$ (for some characteristic examples see Fig. 11). Finally,

Table 10. Intermolecular energies in unpolar systems using pairwise additive potentials

linear chains		$\dfrac{\Delta E^{(n)}\,^a}{m}$
n	$R_e^{(n)}$	
2	1	−1
3	0.9987	−1.0156
4	0.9982	−1.0217
5	0.9980	−1.0248
∞	0.9972	−1.0347
crystals		
hcp	0.97125	−1.4345
fcc	0.97127	−1.4340

[a] m is the number of nearest neighbours

Anton Beyer, Alfred Karpfen and Peter Schuster

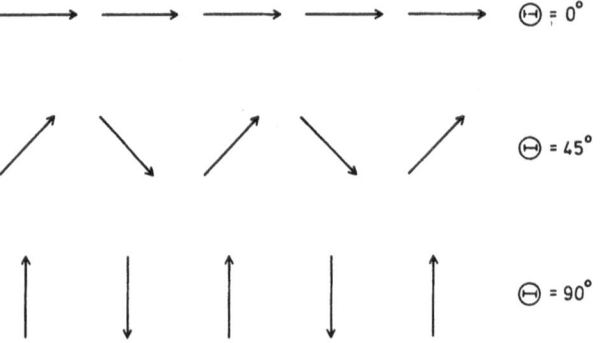

Fig. 11. Arrangements of dipole molecules in periodic chains

we consider the interaction between identical molecules: $\mu_A = \mu_B = \mu$. Eq. (18) then assumes the form

$$\Delta E_{AB} = C_1 \cdot R_{AB}^{-12} - C_2 \cdot R_{AB}^{-6} - \mu^2 (1 + \cos^2 \alpha) R_{AB}^{-3} \qquad (20)$$

In the case of strongly polar systems where the R^{-3} term dominates the attractive part at all relevant values of R. We are dealing again with a two-parameter potential $(C_2 = 0)$ which is defined completely by the equilibrium distance (R_e) and the dissociation energy (D_e):

$$E_{AB} = C_1 \cdot R_{AB}^{-12} - \mu^2 (1 + \cos^2 \alpha) R_{AB}^{-3} = \frac{D_e}{3} \left\{ \left(\frac{R_e}{R_{AB}} \right)^{12} - 4 \left(\frac{R_e}{R_{AB}} \right)^3 \right\}. \qquad (21)$$

Herein we set

$$R_e = \left(\frac{4 C_1}{\mu^2 (1 + \cos^2 \alpha)} \right)^{1/9} \quad \text{and} \quad D_e = \frac{3 \mu^2 (1 + \cos^2 \alpha)}{4 R_e^3}$$

Now, we can apply precisely the same procedure as we did for the Lennard-Jones potential and obtain the corresponding expressions:

$$R_e^{(n)} = R_e \cdot f_n;$$

$$f_n = \left(\frac{n - 1 + (n - 2) \cdot 2^{-12} + (n - 3) \cdot 3^{-12} + \ldots + (n - 1)^{-12}}{n - 1 + (n - 2) \cdot 2^{-3} + (n - 3) \cdot 3^{-3} + \ldots + (n - 1)^{-3}} \right)^{1/9} \qquad (22)$$

$$d_n = \frac{\Delta E^{(n)}(R_e^{(n)})}{(n - 1) D_e} =$$

$$= f_n^{-12} \left(1 + \frac{(n - 2) \cdot 2^{-12}}{n - 1} + \frac{(n - 3) \cdot 3^{-12}}{n - 1} + \ldots + \frac{(n - 1)^{-12}}{n - 1} \right) -$$

$$- 2 f_n^{-3} \left(1 + \frac{(n - 2) \cdot 2^{-3}}{n - 1} + \frac{(n - 3) \cdot 3^{-3}}{n - 1} + \ldots + \frac{(n - 1)^{-3}}{n - 1} \right) \qquad (23)$$

In the limit $n \to \infty$ we obtain for the infinite linear chain of Stockmayer molecules with strong dipoles:

$$R_e^{(\infty)} = R_e \cdot f_\infty; \qquad f_\infty = \left(\frac{\zeta(12)}{\zeta(3)}\right)^{1/9} \tag{24}$$

and

$$d_\infty = \lim_{n \to \infty} \frac{\Delta E^{(n)}(R_e^{(n)})}{(n-1)\,D_e} = -\frac{\zeta(3)^{4/3}}{\zeta(12)^{1/3}}. \tag{25}$$

The expressions for f_n, d_n, f_∞ and d_∞ do not contain parameters and are valid independently of the particular values of C_1, μ or α. To give an example, the further neighbour effects are the same in the three differently oriented chains of dipoles shown in Fig. 11. What changes with α of course, is the nature of R_e and D_e. In Table 10 some numerical results are compiled. As expected, further neighbour effects are more important for polar systems since the R^{-3} potential is further reaching than the R^{-6} potential (see Fig. 10).

In the polar system the contraction of the intermolecular distance is more pronounced: it is about 2% in the infinite chain. The system is stabilized by about 28% of the energy of interaction in the dimer by further neighbour contributions. It should be mentioned that our result neither depends on the absolute value of the dipole moment (μ) nor on the angle formed by the directions of the two dipoles (2α). The only assumption we made concerned the relative weight of R^{-3} and R^{-6} contributions. What do we expect in case none of the two attractive terms is dominant and the potential is to be represented by $-C_2 \cdot R^{-3} = C_3 \cdot R^{-6}$? Basic-

Table 11. Comparison of interatomic or intermolecular distances in vapor-phase dimers and crystals

Monomer	Dimer (vapor phase)		Crystal		
x	R_e (Å)	Ref.	R_e (Å)	Structure[a]	Ref.
He	2.97	[112]	3.0	hcp	[114]
Ne	3.15	[112]	3.156	fcc	[115]
Ar	3.758	[112]	3.755	fcc	[115]
Kr	4.03	[112]	3.992	fcc	[115]
Xe	4.36	[112]	4.335	fcc	[115]
Li	2.673	[112]	3.04	bcc	[116]
			3.09	hcp	[116]
			3.11	fcc	[116]
Na	3.079	[112]	3.72	bcc	[116]
Be[b]	(4.3–4.5)	[113]	2.29	hcp	[116]
Mg	3.891	[112]	3.21	hcp	[116]
HF (x = F)	2.79	[31]	2.49	[c]	[80]
H_2O (x = O)	2.98	[34]	2.74	[c]	[117]

[a] hcp: hexagonal close-packed; bcc: body-centered cubic; fcc: face-centered cubic
[b] The values in parentheses are results of ab initio calculations
[c] see Fig. 12

ally more extensive calculations and a less general result, because the potential can no longer be expressed in terms of D_e and R_e exclusively. From a pragmatic point of view we can be sure, however, that the ultimate result will lie somewhere in between the bounds determined by the 12-6-(Lennard-Jones) or the 12-3-potential. The stronger effect observed with the polar systems has its origin in the longer effective range of dipole forces expressed by the R^{-3} dependence of the attractive part of the energy of interaction which always exceeds the R^{-6} contribution of the dispersion energy at sufficiently large intermolecular distances.

Let us now confront the ideas of a world of pairwise additive potentials with reality, in particular with the actual experimental data on structures. To give a straightforward description of the nature of further neighbour interactions we compare intermolecular distances in the vapour-phase dimers and in crystals with simple lattice structures. In the case of solids formed by noble gas atoms we find very small contractions indeed (Table 11). The differences in the equilibrium distances between vapor-phase and crystal is a few hundreths of an Angstrom or even less. Nevertheless, these minute energetic effects have been debated extensively because they have a characteristic influence on the observed crystal structures: Thus, instead of the hexagonal close-packed (hcp) structure the face-centered cubic lattice (fcc) is found to be the most stable arrangement [53]. The only exception is solid helium which prefers the expected hcp lattice structure. Whether this phenomenon can be attributed to the operation of three-body forces or is the result of some other higher order effect is not yet completely clear.

Metal atoms with a simple electronic structure show very different effects when we compare vapor-phase dimers and crystals. The differences are large and of opposite sign in the case of alkali metals and alkaline earth metals. The dimers of alkaline metal atoms in the gas phase, Li_2, Na_2 etc., are characterized by a weak covalent

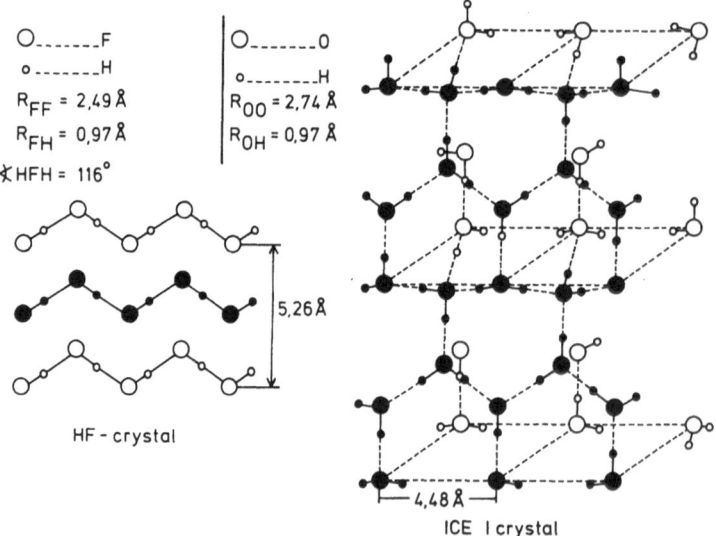

Fig. 12. Crystal structures of HF and H_2O

bond which is markedly stretched in the solid state. Hence, we are dealing with a repulsive contribution of many-body forces. Alkaline earth metals behave completely differently: Mg_2 is a typical "van der Waals molecule", similar to the noble gas dimers Ne_2 or Ar_2. The data on Be_2 as obtained by ab initio calculations are highly uncertain [54-56]. Attempts to prove the existence of a bound ground state in the vapor phase by spectroscopic techniques have failed so far. Nevertheless, it is certain that in the case of Be and Mg the equilibrium distance in the crystal is substantially smaller than that in the vapor-phase dimer. In both systems stabilization occurs through three-body forces. For a more detailed discussion see [57,58].

Let us now consider systems formed by polar molecules, e.g. HF, H_2O and HCl. The HF and HCl crystals contain one-dimensional bent chains of molecules between which the mutual interactions are relatively weak (Fig. 12). In the case of HF we observe a marked decrease of the intermolecular distance ($\Delta R_{FF} \sim 0.3$ Å) upon the formation of the solid phase. Ice I has a fairly complicated three-dimensional structure (Fig. 12), dipoles appear at different relative orientations, and the infinite chain is no appropriate model. Nevertheless, the contraction of the intermolecular distance in the solid state is substantial ($\Delta R_{OO} \sim 0.24$ Å). In both cases, the stabilizing contributions have to be attributed to attractive many-body forces since the changes observed exceed by far the effects to be expected in polar systems with pairwise additive potentials. The same is true for the energy of interaction (Table 12):

Table 12. Comparison of intermolecular distances and hydrogen-bond energies in vapor-phase dimers and crystals

Vapor phase dimer				Crystal				
	R_{xx} (Å)	E (kJ/mol)	Ref.	R_{xx} (Å)	E (kJ/mol)	Ref.	f_∞	d_∞
HF (calc.)	2.830	−15.90	[30]	2.598	−27.62	[87]	0.918	1.737
(exp.)	2.789		[31]	2.498		[80]	0.896	
HCl (calc.)	3.757	−7.95	[71]	3.731	−9.21	[71]	0.993	1.158
(exp.)				3.710		[88]		
H_2O (calc.)	2.99	−18.00	[14]					
(exp.)	2.976	−22.60	[34, 90]	2.74	−29.47	[117]	0.920	1.304

the factor d for the infinite HF chain, $d_\infty = 1.711$, is much higher than the limiting value for pairwise additive polar systems ($d_\infty = 1.278$). This also applies to Ice I.

A very interesting case is hydrogen chloride. The vapor-phase dimer has been studied extensively by accurate ab initio calculations [59]. We can regard these results as highly relaible. No direct experimental information on $(HCl)_2$ in the vapor phase is available at present. Some matrix isolation data are discussed in Section 4.3. The crystal structure of HCl is closely related to that of HF (Fig. 12). The infinite bent chains are relatively loosely packed, and the system is a good test case for pairwise additivity. Indeed, we observe a certain contraction of the intermolecular distance in the crystal relative to the dimer. This contraction, however, is much smaller than that in the HF crystal and falls in the range between the contractions typical of unpolar and polar crystals with pairwise additive potentials ($f_\infty = 0.993$).

We may conclude that many-body forces are not important for the structure of solid hydrogen chloride (for further details see Sections 4.3 and 5). The energy of interaction in the dimer and in the solid fit very well into our relations. This is more a test of our assumptions of binary potentials in equations 8 and 18 than a limit on the role of many-body forces because the only available value was derived from cluster calculations based on the assumption of pairwise additivity. From the concepts and data discussed in this section it is obvious that an accurate description of clusters and condensed phases formed from polar molecules like HF and H_2O which are both characteristic hydrogen bond donors *and* acceptors, requires a proper consideration of many-body forces.

4.2 Trimers of Hydrogen Fluoride and Water

After this excursion into a world of pairwise additive potentials we summarize the available data on $(HF)_3$ and $(H_2O)_3$. We start with the results of ab initio calculations. The most stable configurations of the hydrogen-bonded trimers $(HF)_3$ and $(H_2O)_3$ are cyclic structures (Fig. 13). Because the enormous numerical efforts which are inevitable in large scale computations on trimers no satisfactory equilibrium geometries are available at present. Most calculations were performed assuming frozen monomer geometries or applying some other constraints.

The most extensive study on $(HF)_3$ has been performed recently [60]. Full geometry optimization of the cyclic trimer was carried out with the constraint of C_{3h} symmetry. Earlier calculations using small basis sets and frozen monomer geometries [61] led to a most stable planar structure exhibiting this symmetry. As expected from ring strain, the deviation from linearity of the hydrogen bond is rather large ($\theta = 26.9°$). The intermolecular distance is smaller than in the dimer: $R_{FF} = 2.71$ Å compared to $R_{FF} = 2.83$ Å. The HF bond length shows more elongation than in the dimer: $\Delta R_{HF} = 0.010$ Å.

In the case of the water trimer the situation is much more complex. An extensive study has been performed with frozen monomer geometries and under the assumption of pairwise additive potentials [20]. This structure (Fig. 13) is characterized by an almost planar arrangement of the tree oxygen and the three hydrogen atoms

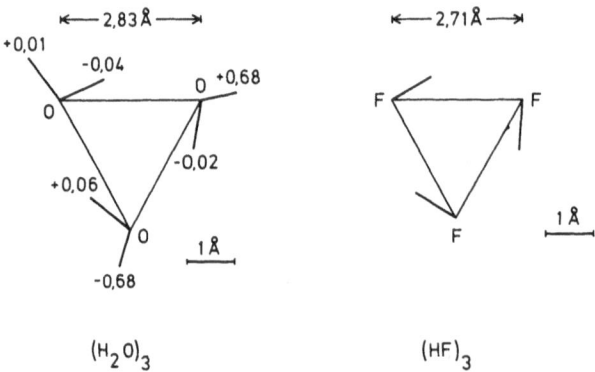

Fig. 13. Computed equilibrium structures of cyclic $(H_2O)_3$ [73] and $(HF)_3$ [60]

involved in hydrogen bonding. This fragment shows roughly C_{3h} symmetry. The remaining three hydrogen atoms point out from the ring in different directions: one is almost in the plane of the ring, one above and one below.

Direct structural information on HF oligomers in the vapor phase is confined to the hexamer: Janzen and Bartell [62] proved the existence of cyclic hexamers by means of an elegant combination of molecular beam and electron diffraction techniques. In contrast, in the case of water oligomers only indirect information is available. Aggregates of H_2O molecules were experimentally studied with the "molecular-beam electric resonance" technique [63]. Higher aggregates were formed on the expansion of the beam from the supersonic nozzle. The molecular weight of the clusters was determined by mass spectrometry. These studies reveal that the higher water aggregates cannot be focussed in the "electric-resonance" spectrometer and hence are unpolar. The absence of a permanent electric dipole moment in these aggregates is an indication of cyclic structures.

Open-chain trimers of hydrogen fluoride and water have been studied to a certain extent because these systems occur as structural fragments in solid hydrogen fluoride and ice (Fig. 12). In addition, they are of particular interest for studies on hydrogen bond additivity. The determination of equilibrium structures of open-chain trimers by energy oprimization is much more demanding than in case of the cyclic conformations. Since no symmetries can be presumed the number of geometric parameters to be varied is very large. No complete optimizations have been performed so far.

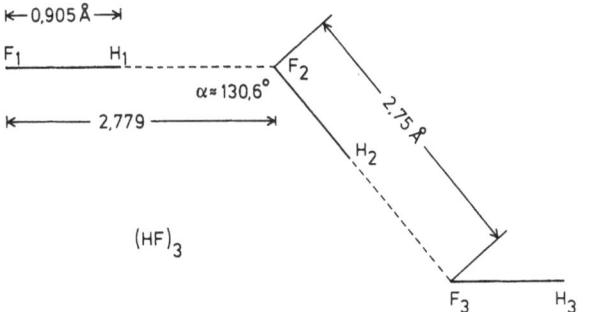

Fig. 14. Computed equilibrium structure of open-chain (HF)$_3$ [60]

The approximate equilibrium geometry of the open-chain structure of (HF)$_3$ is shown in Fig. 14. The energy of interaction relative to three HF molecules is -9.4 kcal/mol. The open-chain trimer structure is thus less stable than the cyclic trimer by approximately 2.2 kcal/mol. For the former structure we calculate an energy per hydrogen bond of -4.7 kcal/mol whereas for the latter we find only a value of -3.9 kcal/mol. Thus, the ring strain in the cyclic structure reduces the energy of interaction. The cyclic conformer, however, contains one more bond which ultimately leads to the higher stability.

Hydrogen fluoride trimer has also been used as an appropriate model system for the discussion of three-body forces [64,65]. The data reported on (HF)$_3$ indicate a stabilizing effect caused by the three-body interactions. The energy per hydrogen

Table 13. Computed equilibrium structures of cyclic and open-chain hydrogen fluoride trimers

Cyclic trimer

Basis	R_{FF} (Å)	R_{HF} (Å)	α (°)	E (kJ/mol)	Ref.
A STO-4G	2.31	0.938[a]	20	−81.0	61)
B (10, 5, 1/5, 1) → [5, 3, 1/3, 1]	2.72	0.910	26.7	−60.6	60)
C (10, 6, 2/6, 1) → [6, 4, 2/4, 1]	2.73	0.909	26.9	−48.5	60)

[a] constrained value

Open-chain trimers[b]

Basis	$R_{F_1F_2}$ (Å)	$R_{F_2F_3}$ (Å)	$R_{F_1H_1}$ (Å)	$R_{F_2H_2}$ (Å)	$R_{F_3H_3}$ (Å)	$\measuredangle F_1F_2H_3$ (°)	$\measuredangle F_2F_3H_3$ (°)	E (kJ/mol)	
A	2.55[a]	2.55[a]	0.938[a]	0.938[a]	0.938[a]	107[a]	111[a]	−55.1	cis (minimum geometry of dimer)
A	2.46	2.46	0.938[a]	0.938[a]	0.938[a]	107[a]	111[a]	−56.1	cis, R_{FF} not fixed
A	2.55[a]	2.55[a]	0.938[a]	0.938[a]	0.938[a]	107[a]	111[a]	−53.8	trans (minimum geometry of dimer)
B	2.75	2.72	0.905	0.907	0.903	132.0	133.5	−48.6	
C	2.78	2.75	0.905	0.907	0.903	130.6	130.6	−39.3	

[a] constrained value; [b] for the definition of internal coordinates see Fig. 14

bond is therefore substantially larger than in the dimer. A quantitative analysis is given in Table 13. The additional stabilization of the trimer can be interpreted as follows: in the open-chain trimer the interaction of the central HF molecule with one neighbour leads to a change in the electron density distribution which enhances the strength of the hydrogen bond to the other neighbouring HF molecule and vice versa. Thus, the increase in polarity of molecules involved in hydrogen bonding goes hand in hand with the non-additivity of intermolecular interactions originating from polarization.

In a recent thermodynamic and spectroscopic study Redington [66–68] evaluated ΔH and ΔS for the formation of cyclic and open-chain oligomers of hydrogen fluoride. His best fit to vapor density, heat capacity, excess entropy, excess enthalpy, and infra red absorption data shows a monotonous increase in ΔH per hydrogen bond with increasing number of HF molecules both for open-chain and cyclic clusters, using the relation

$$\Delta H/\text{bond} = \Delta H_{(HF)_n}/k \qquad (26)$$

for the determination of ΔH per hydrogen bond.

For cyclic clusters $k = n$, for open-chain clusters $k = n - 1$. This behaviour was imposed on the fitting procedure by the model assumptions applied. In the case of open-chain clusters this regular increase agrees with quantum chemical studies. Applying Eq. (26), $\Delta H/\text{bond}$ values are extremely slowly converging with respect to n because each hydrogen bond in the cluster enters with equal statistical weight. An alternative and much faster converging way [69] to obtain the limiting value of ΔH per hydrogen bond for large n is to use the relation

$$\Delta H/\text{bond} = \Delta H_{(HF)_n} - \Delta H_{(HF)_{n-1}} \qquad (27)$$

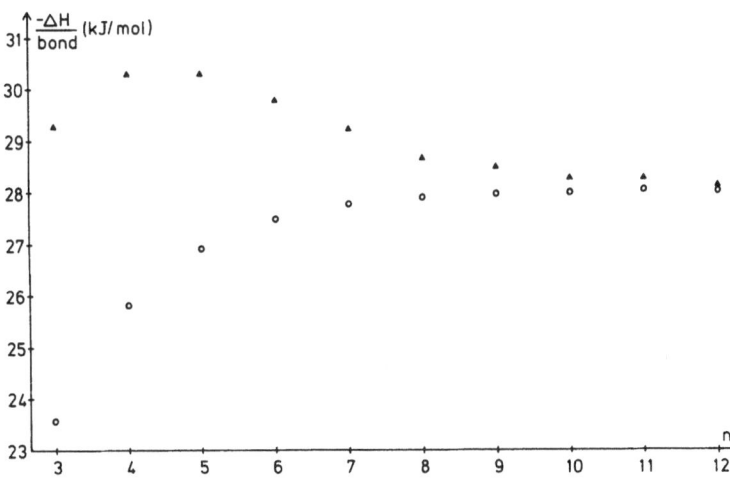

Fig. 15. ΔH/bond of open-chain clusters of increasing length calculated from Eqs. (26) ($\bigcirc$) and (27) ($\triangle$). ΔH values for clusters are taken from ref. [67]

Edge effects approximately cancel each other and the application of Eq. (27) corresponds to the insertion of an additional HF molecule into the center of the chain. Using the thermodynamic parameters given by Redington, ΔH/bond values of open-chain oligomers are calculated from Eqs. (26) and (27) and plotted in Fig. 15 as a function of n. According to expectations, initially a much sharper increase of ΔH/bond is observed if the data are calculated from Eq. (27). A small but significant maximum is, however, observed around n = 4 or 5. Obviously, both curves converge to the same limiting value. One may therefore conclude that either the limiting value given by Redington is too low by about 0.5 kcal/mol or that the stability of the smaller clusters is slightly overestimated. Notwithstanding the practical difficulties, it would seem preferable to use both Eq. (26) and (27) as constraints to thermodynamic fitting parameters in the case of open-chain oligomers.

For the case of cyclic oligomers the situation is certainly more complicated. As a consequence of a preferred angle α_{dimer} between the molecular axes in the dimer the use of Eq. (27) should lead to a maximum at a certain critical ring size coinciding approximately with n = 360/(180 − α_{dimer}). Both in the case of smaller and larger rings ring closure must result in energetically disadvantageous configurations as a consequence of ring strain. In the case of hydrogen fluoride this explains the greater stability of the cyclic hexamer. Analogous considerations for hydrogen chloride lead to a preference for cyclic tetramers because α_{dimer} is close to 90°. Both matrix isolation studies [70] and model calculations [71] indeed show a maximum in the stabilization energy for cyclic HCl tetramers.

Two systematic studies based on extensive ab initio calculations of water trimer with open-chain configuration have been reported so far [72,73]. In these studies some characteristic geometrical arrangements were chosen to investigate the role of

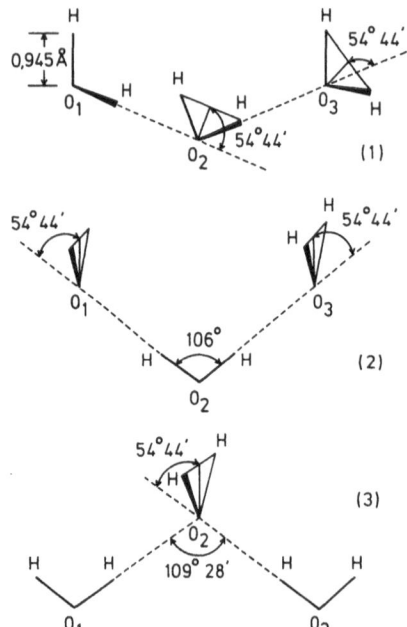

Fig. 16. Geometries of some selected open-chain water trimers [72]

three-body forces. Depending on the relative orientation of the two hydrogen bonds in the trimer, we find stabilizing or destabilizing three-body interactions. In the sequential trimer (Fig. 16) the hydrogen bonds are in sequence as they were in the structure of $(HF)_3$ considered above. The stabilizing three-body term in this configuration is of similar magnitude as that found in the analogous hydrogen fluoride complex. In the double-donor and double-acceptor trimer three-body forces are repulsive. Polarization in these cases leads to mutual weakening of the hydrogen bonds. It is interesting to note that all the trimer subunits in solid hydrogen fluoride are of the sequential type and hence we expect optimal mutual reinforcement of individual hydrogen bonds (for infinite chains see also section 4.3). In ice I we observe all three kinds of trimer substructures, sequential, double donor and double acceptor. Hence, there occurs partial compensation of three-body interactions, and the overall effects in the crystal, e.g. the decrease in the intermolecular distance between vapor-phase dimer and crystal, are expected to be weaker in ice than in the hydrogen fluoride crystal (cf. Table 11). This is found indeed experimentally.

4.3 Hydrogen Bonding in Infinite Chains

Theoretical studies on hydrogen bonded complexes, dimers or oligomers, although tedious and time-consuming if accurate results are to be achieved, are nowadays quite standard due to the worldwide distribution of efficient program packages. Extrapolation of oligomer properties to the properties of molecules embedded in infinite chains is, however, hampered by the slow convergence of various expectation values with respect to chain length. Moreover, the enormous number of internal degrees of freedom does not allow complete optimization of equilibrium geometries in large clusters. Explicit use of translational symmetry is therefore considerably more efficient in order to obtain these desired limiting values which, depending on the actual structure of hydrogen-bonded molecular crystals may either furnish a direct comparison with solid state data or at least allow some important consequences of the non-additive behaviour of model systems to be studied.

The formalism to incorporate translational symmetry into the usual Hartree-Fock approach, the crystal orbital technique, is not new at all [74, 75]. Reviews of recent developements and applications of the Hartree-Fock crystal orbital method may be found in refs. [76-79]. However, only few investigations on the evaluation of equilibrium geometries and other properties derived from computed potential surfaces of one-dimensional infinite crystals or polymers have been reported.

Many molecular crystals held together by hydrogen bonds form catameric, chain-like patterns as a constituting structural element in the solid state. We mention here only formic acid, acetic acid, simple alcohols such as methanol and ethanol, hydrogen cyanide, and hydrogen halides. From the latter group, hydrogen fluoride is best suited for theoretical analysis. Hydrogen fluoride crystals have been studied by a variety of spectroscopic methods. Thus, their structure has been determined both by X-ray [80] and neutron diffraction [81] measurements and their vibrational spectra have been studied by infra red [82], Raman [83] and neutron spectroscopy [84]. NMR investigations have been reported as well [85].

Structural data are the most important ones for our purpose here. As discussed in section 4.1 the additive model of intermolecular interactions fails in the case of

Table 14. Comparison of computed equilibrium structures and hydrogen-bond energies of HF, $(HF)_2$ and $(HF)_\infty$

	R_{HF} (Å)	R_{FF} (Å)	$\not\prec$ FFF (°)	ΔE (kJ/mol)	Ref.
HF calc.	0.900	—	—	—	[105]
exp.	0.917	—	—	—	[105]
$(HF)_2$ calc.	0.904	2.83 ·	123.2	−16.7	[30]
exp.	—	2.79	108.0	—	[31]
$(HF)_\infty$ calc.	0.918	2.60	129.7	−27.2	[87]
exp.	0.97	2.49	116.0	—	[80]

hydrogen fluoride chains. Theoretical descriptions of this system, therefore, have to go beyond the approximation of pairwise additivity. Linear models for the actually bent hydrogen fluoride chains were investigated earlier [32, 86] (for a detailed treatment of the bent chain see ref. [87]). Results on the equilibrium structure and hydrogen bond energy are collected in Table 14.

The reduction of the intermolecular distance is significant in the infinite chain. A large percentage ($\sim 80\%$) of the experimentally determined shortening is reproduced at the Hartree-Fock level. The elongation of the intermolecular bond, although distinctly larger than in the dimer, is underestimated in our calculations. It is still an open question whether this discrepancy is to be attributed to the lack of electron correlation contributions or to a consequence of the harmonic approximation. The most striking feature of cooperativity originating from the regular lattice structure is a radical change in the potential surface for simultaneous proton motion. An appropriate shift of the proton sublattice conserving the helical structure follows a symmetric double minimum potential. Since this motion roughly corresponds to the spectroscopically active symmetric stretching mode, the vibrational frequency is drastically reduced by about $1000 \, \text{cm}^{-1}$ compared to the monomer frequency in the vapor phase. This frequency shift is larger by more than a factor of ten than the frequency shift in the dimer. As a consequence of the much smaller intermolecular distance, all intermolecular force constants and hence also vibrational frequencies are considerably increased, notably those for bending vibrations. Table 15 gives a comparison of computed vibrational frequencies in HF, $(HF)_2$ and $(HF)_\infty$. Considering the substantial changes on polymer formation we realize again the inappropriateness of additive models relying inherently on frozen intramolecular geometries. Another important quantity for the characterization of non-additive contributions is the change in electric moments per molecule in the process of polymer formation. Particular care has to be exercised in case these quantities are extracted from cluster calculations. Dividing the total dipole moment of an open-chain cluster by the number of molecules and subsequent extrapolation to large n does not yield the desired result. Because of charge transfer the molecules at both ends of the cluster carry partial charges. Electric neutrality is only approached for the central molecules in larger chains. Therefore, much too large dipole moments per molecule are obtained if the above mentioned procedure is chosen. A more correct procedure involves a partitioning of the electronic contributions to the dipole moment similar to a Mulliken population analysis collecting all one-center dipole integrals whereby at least one basis function stems from the central molecule. Extrapolation of this quan-

Table 15. Computed harmonic vibrational frequencies of HF, $(HF)_2$ and $(HF)_\infty$ using a $(9, 5, 1/5, 1) \rightarrow [5, 3, 1/3, 1]$ basis set. All values in cm^{-1}

	vibrational mode		frequency
HF	intramolecular		4494
$(HF)_2$	intramolecular		4440, 4396
	intermolecular	in-plane	528, 209, 156
		out-of-plane	436[a]
$(HF)_\infty$[b]	intramolecular		4170, 3967
	intermolecular		
	in-plane bending		1007, 604
	out-of-plane bending		716, 563
	translations		386, 150

[a] value taken from ref. [30]; [b] ref. [87]

tity to large n exactly leads to the dipole moment per molecule obtained from crystal orbital calculations. From [87] we infer an increase of about 20 % of the dipole moment of a hydrogen fluoride molecule embedded in a chain compared to an isolated HF molecule in the vapor phase. Again direct computation of this quantity using the crystal orbital approach is by far more efficient, due to the slow decay of edge effects in open-chain clusters.

The crystal structure of hydrogen chloride resembles that of hydrogen fluoride [88]. Planar bent chains are formed with an angle of about 90° between neighbouring molecules. Thus the Cl—Cl—Cl angle is significantly smaller then the F—F—F angle in the analogous chain. Most probably, this is a consequence of the much larger quadrupole moment and much smaller dipole moment of HCl compared to those of the HF molecule. The almost negligible reduction of the intermolecular distance in the hydrogen chloride chain is a direct result of the much weaker dipole-dipole contribution to the intermolecular energy. Since quadrupole-quadrupole interactions decay with R^{-5}, long-range interactions are less important too. Although, as discussed in [59], Hartree-Fock calculations on $(HCl)_2$ still predict the correct intermolecular orientation, the dominating contribution to the intermolecular interaction is the dispersion energy. Polarization non-additivities are therefore not important in the case of solid HCl. HCl clusters have been investigated with the aid of matrix isolation spectroscopy. A distinct preference for cyclic trimers and tetramers was observed. Higher aggregates were not detected.

5 What is a Hydrogen Bond?

A large number of hydrogen-bonded complexes has been studied so far. Due to extensive ab initio calculations and careful spectroscopic studies some of them are as well known in the vapor phase as are molecules of comparable size. As far as clusters of molecules are concerned the present knowledge is rather fragmentary. This lack reflects the enormous difficulties which arise both in computational and spectroscopic studies on systems with many degrees of freedom. Large amplitude motions as they occur in most of these complexes make the analysis even more difficult. The results obtained so far nevertheless allow some general conclusions to be made. The strength of a hydrogen bond correlates with the basicity of the proton-acceptor and the acidity of the proton-donor molecule. These correlations are well established for the first row atoms N, O and F involved in hydrogen bonding: the energy of interaction increases e.g. in the two series of complexes HF ... HF, H_2O ... HF, H_3N ... HF and H_3N ... HNH_2, H_3N ... HOH, H_3N ... HF.

Hydrogen bonds are usually characterized by the following structural and spectroscopic features.

1. The distances between the neighbouring atoms of the two functional groups (AH ... B) involved in the hydrogen bond are substantially smaller than the sum of their van der Waals' radii: $R_{HB} < R_H^0 + R_B^0$.

2. The AH bond length is increased upon hydrogen bond formation

3. The AH bond stretching modes are shifted towards lower frequencies on hydrogen bond formation.

4. The polarities of AH bonds increase upon hydrogen bond formation, usually leading to larger dipole moments of the complexes than expected from vectorial addition of the components. Furthermore, also the significantly enhanced IR intensities indicate an increase in the dipole moment derivatives on hydrogen bond formation.

5. NMR chemical shifts of protons in hydrogen bonds are substantially smaller than those observed in the corresponding isolated molecules. The observed deshielding effect is due to the reduced electron densities at protons involved in hydrogen bonding.

In addition to these commonly accepted characteristics of hydrogen bonding we would like to guide the readers attention to a sixth regularity which allows to confine the very general concept of a hydrogen bond to the more typical cases. In the case of vapor-phase dimers the effects described in (1) to (4) are very weak compared to the corresponding effects occurring in larger clusters and condensed phases. This fundamental difference between vapor-phase dimers and higher aggregates is a measure of the mutual enhancement through polarization. It reflects the non-additive

Table 16. Wave numbers (cm^{-1}) of HX stretching vibrations in the gaseous and solid state

	Monomer, vapor phase		Dimer, vapor phase		Infinite chain, crystal	
	HX	Ref.	$(HX)_2$	Ref.	(HX)	Ref.
H—F	3962	[109]	3895	[109]	3583	[82]
			3857		3404	
					3275	
					3065	
H—Cl	2888	[112]	$(2856)^a$	[70]	2758	[118]
	$(2869)^a$	[70]	$(2818)^a$		2748	
					2720	
					2705	
H—Br	2559	[112]	$(2550)^a$	[70]	2440	[118]
	$(2556)^a$	[70]	$(2496)^a$		2431	
					2406	
					2395	

[a] measurements performed in Ar matrix at low temperature

contribution to intermolecular forces which are basically represented by three-body terms. When the differences between the vapor phase dimer and condensed state are large we can conclude that three- and higher many-body polarization effects occur. This feature is characteristic of hydrogen-bonded networks in which the lone-pair donors (B) are first row atoms and the AH bond contains a first row atom as well. We illustrate this conjecture by the comparison shown in Table 16.

Vibrational frequencies were chosen because they are very sensitive to changes in bond strengths. The hydrogen halides HF, HCl and HBr are appropriate because they have simple structures with all dipoles aligned sequentially. True vapor-phase data are available only for hydrogen fluoride. Concerning the two other systems we have to rely on the results of low-temperature isolation studies in Ar matrices. However, the differences between the vibrational frequencies of the hydrogen halides in the vapor phase and those measured in the matrix are so small that they do not affect at all our conclusions. The difference in the frequency shifts between HF on the one hand and HCl and HBr on the other hand is remarkable: in the former case the shifts are larger by a factor of five. We now ask what are the differences in molecular properties which cause such different properties of structurally close related crystals. The acidity of the hydrogen halides increases in the series HF, HCl, HBr and so does also their capability to form hydrogen bonds as donor molecule. A comparison of the energies of interaction in complexes of the same base with HF and HCl shows that the latter forms somewhat stronger hydrogen bonds. Thus the basic difference between HF and HCl is that HCl is a much weaker hydrogen-bond acceptor than HF. Hence, the hydrogen bond in $(HCl)_2$ is rather weak. Moreover, the energy of interaction of this complex is dominated by the dispersion term [59]: $\Delta E_{DIS} \sim 0.6 \, \Delta E$. Mutual polarization is therefore much less important than in aggregates of HF, and hence the relative contribution of three-body terms to the structures and properties of $(HCl)_n$ clusters and liquid and solid hydrogen chloride are negligibly small. However, the characteristic properties of associated liquids, such as water and hydrogen fluoride, and of the corresponding solids cannot be understood completely without explicit consideration of the deviations from pairwise additive potentials.

6 Acknowledgments

We thank Prof. R. Ahlrichs for communicating results prior to publication, Dr P. Herzig and Mag. Zobetz for their help in lattice-summation calculations, and Mrs. J. Jakubetz and Mr. J. König for the preparation of the manuscript. Financial assistance by the Austrian "Fonds zur Förderung der wissenschaftlichen Forschung" (Project No. 3388 and 3669) and the generous supply with computer time by the Interfakultäres Rechenzentrum, Wien, are also gratefully acknowledged.

7 References

1. The Hydrogen Bond — Recent Developments in Theory and Experiments, Vols. I–III. (eds.) Schuster, P., Zundel, G., Sandorfy, C., Amsterdam, North Holland 1976
2. Schuster, P.: Intermolecular Interactions from Diatomics to Biopolymers (ed.) Pullman, B., p. 363, Chichester, Wiley 1978
3. Kollman, P.: J. Am. Chem. Soc. 99, 4875 (1977)

4. Kollman, P.: Applications of electronic structure theory, in: Modern Theoretical Chemistry (ed.) Schaefer III, H. F., Vol. 4, p. 109, New York, Plenum Press 1977

5. Schuster, P.: Angew. Chem. Internat. Ed. Engl. 20, 546 (1981)

6. Zundel, G.: in ref. 1, Vol. II, p. 683

7. Olovsson, I., Jönsson, P. G.: in ref. 1, Vol. II, p. 393

8. Kollman, P. A., Johansson, A., Rothenberg, S.: Chem. Phys. Lett. 24, 199 (1974)

9. Methods of electronic structure theory, in: Modern Theoretical Chemistry (ed.) Schaefer III, H. F., Vol. 3, New York, Plenum Press 1977

10. Maitland, G. C., Rigby, M., Smith, E. B., Wakeham, W. A.: Intermolecular Forces, Oxford, Clarendon Press 1981 .

11. Ahlrichs, R., Penco, P., Scoles, G.: Chem. Phys. 19, 119 (1977)

12. a) Klemperer, W.: Faraday Disc. 62, 179 (1977)
    b) Klemperer, W.: J. Mol. Struct. 59, 161 (1980)

13. Boys, S. F., Bernardi, F.: Mol. Phys. 19, 558 (1970)

14. Johansson, A., Kollman, P., Rothenberg, S.: Theoret. Chim. Acta 29, 167 (1973)

15. Jeziorsky, B., Van Hemert, M.: Mol. Phys. 31, 713 (1976)

16. Brzić, A., Karpfen, A., Lischka, H., Schuster, P.: in preparation

17. Denisov, G. S.: Lecture at the 6th European Workshop on Horizons in Hydrogen Bond Research, Netherlands, Leuwen 1982

18. Yarkony, D. R., O'Neil, S. V., Schaefer III, H. F., Baskin, C. P. Bender, C. F.: J. Chem. Phys. 60, 855 (1974)

19. Popkie, H., Kistenmacher, H., Clementi, E.: J. Chem. Phys. 59, 1325 (1973)

20. Kistenmacher, H., Lie, G. C., Popkie, H. and Clementi, E.: J. Chem. Phys. 61, 546 (1974)

21. Matsuoka, O., Clementi, E., Yoshimine, M.: J. Chem. Phys. 64, 1351 (1976)

22. Curtiss, L. A., Pople, J. A.: J. Mol. Spectry. 61, 1 (1976)

23. Curtiss, L. A., Pople, J. A.: J. Mol. Spectry. 55, 1 (1975)

24. Hobza, P., Zahradnik, R.: Top. Current Chem. 93, 53 (1980)

25. Hobza, P., Zahradnik, R.: Chem. Phys. Lett. 82, 473 (1981)

26. Bouteiller, Y., Guissani, Y.: Chem. Phys. Lett. 69, 280 (1980)

27. Bouteiller, Y., Allavena, M., Leclercq, J. M.: Chem. Phys. Lett. 84, 363 (1981)

28. Bouteiller, Y., Allavena, M., Leclercq, J. M.: J. Chem. Phys. 73, 2851 (1980)

29. Newton, M. D., Kestner, N. R.: Chem. Phys. Lett. 94, 198 (1983)

30. Lischka, H.: Chem. Phys. Lett. 66, 108 (1979)

31. Dyke, T. R., Howard, B. J., Klemperer, W.: J. Chem. Phys. 56, 2442 (1972)

32. Karpfen, A., Schuster, P.: Chem. Phys. Lett. 44, 459 (1976)

33. Dyke, T. R., Mack, K. M., Muenter, J. S.: J. Chem. Phys. 66, 498 (1977)

34. Odutola, J. A., Dyke, T. R.: J. Chem. Phys. 72, 5062 (1980)

35. Dyke, T. R.: J. Chem. Phys. 66, 492 (1977)

36. Slanina, Z.: Adv. Mol. Rel. Int. Proc. 19, 117 (1981)

37. Braun, C., Leidecker, H.: J. Chem. Phys. 61, 3104 (1974)

38. Gebbie, H. A., Bohlander, R. A., Pardoe, G. W. F.: Nature 230, 521 (1971)

39. Ayers, G. P., Pullin, A. D. E.: Spectrochim. Acta 32A, 1629, 1641, 1689, 1695 (1976)

40. Fredin, L., Nelander, B., Ribbegard, G.: J. Chem. Phys. 66, 4065 (1977)

41. Hagen, W., Tielens, A. G. G. M.: J. Chem. Phys. 75, 4198 (1975)

42. Barnes, A. J., Szczepaniak, K., Orville-Thomas, W. J.: J. Mol. Struct. 59, 39 (1980)

43. Behrens, A., Luck, W. A. P.: J. Mol. Struct. 60, 337 (1980)

44. Sokalsky, W. A.: J. Chem. Phys. 77, 4529 (1982)

45. Kisiel, Z., Legon, A. C., Millen, D. J.: Proc. Roy. Soc. (London) A381, 419 (1982)

46. Bevan, J. W., Legon, A. C., Millen, D. J., Rogers, S. C.: Chem. Commun. 1975, 341

47. Bevan, J. W., Kisiel, Z., Legon, A. C., Millen, D. J., Rogers, S. C.: Proc. Roy. Soc. (London) A372, 441 (1980)

48. Legon, A. C., Willoughby, L. C.: Chem. Phys. Lett. 92, 333 (1982)

49. a) Keenan, M. R., Buxton, L. W., Campbell, E. J., Legon, A. C., Flygare, W. H.: J. Chem. Phys. 74, 2133 (1981)
    b) Dixon, T. A., Joyner, C. H., Baiocchi, F. A., Klemperer, W.: J. Chem. Phys. 74, 6539 (1981)

50. Buxton, L. W., Campbell, E. J., Keenan, M. R., Balle, T. J., Flygare, W. H.: Chem. Phys. 54, 173 (1981)

51. Thomas, R. K.: Proc. Roy. Soc. (London) *A344*, 579 (1975)
52. Bevan, J. W., Martineau, B., Sandorfy, C.: Can. J. Chem. *57*, 1341 (1979)
53. Niebel, K. F., Venables, J. A.: in: Rare Gas Solids (eds.) Klein, M. L., Venables, J. A., Vol. 1, p. 558, New York, Academic Press 1976
54. Chiles, R. A., Dykstra, C. E.: J. Chem. Phys. *74*, 4545 (1981)
55. Blomberg, M. R. A., Siegbahn, P. E. M., Roos, B. O.: Intern. J. Quant. Chem. *S14*, 229 (1980)
56. Liu, B., McLean, A. D.: J. Chem. Phys. *72*, 3418 (1980)
57. Kolos, W.: New Horizons of Quantum Chemistry (eds.) Löwdin, P. O., Pullman, B., p. 243, Reidel 1983
58. Murrell, J. N.: Chem. Phys. Lett. *55*, 1 (1978)
59. Votava, C., Ahlrichs, R. in: Intermolecular Forces, Proc. of the 14th Jerusalem Symp. (ed.) Pullman, B., Reidel 1981
60. Karpfen, A., Beyer, A., Schuster, P.: to be published
61. Del Bene, J. E., Pople, J. A.: J. Chem. Phys. *55*, 2296 (1971)
62. Janzen, J., Bartell, L. S.: J. Chem. Phys. *50*, 3611 (1969)
63. Dyke, T. R., Muenter, J. S.: J. Chem. Phys. *57*, 5011 (1972)
64. Beyer, A., Karpfen, A., Schuster, P.: Chem. Phys. Lett. *67*, 369 (1979)
65. Schuster, P., Karpfen, A., Beyer, A., in: Molecular Interactions (eds.) Ratajczak, H., Orville-Thomas, W. J., p. 117, New York, Wiley 1980
66. Redington, R. L.: J. Chem. Phys. *75*, 4417 (1981)
67. Redington, R. L.: J. Phys. Chem. *86*, 552 (1982)
68. Redington, R. L.: J. Phys. Chem. *86*, 561 (1982)
69. Karpfen, A., Ladik, J., Russegger, P., Schuster, P., Suhai, S.: Theoret. Chim. Acta *34*, 115 (1974)
70. Maillard, D., Schriver, A., Perchard, J. P., Girardet, C.: J. Chem. Phys. *71*, 505 (1979), ibid. *71*, 517 (1979)
71. Votava, C., Ahlrichs, R., Geiger, A.: to be published in J. Chem. Phys.
72. a) Hankins, D., Moskowitz, J. W., Stillinger, F. H.: Chem. Phys. Lett. *4*, 527 (1970);
    b) Hankins, D., Moskowitz, J. W., Stillinger, F. H.: J. Chem. Phys. *53*, 4544 (1970)
73. Clementi, E., Kolos, W., Lie, G. C., Ranghino, G.: Intern. J. Quant. Chem. *17*, 377 (1980)
74. DelRe, G., Ladik, J., Biczo, G.: Phys. Rev. *155*, 997 (1967)
75. André, J. M., Gouverneur, L., Leroi, G.: Intern. J. Quant. Chem. *1*, 451 (1967)
76. Ladik, J., Suhai, S.: Specialists Periodic Reports on Theoretical Chemistry (ed.) Vol. 4, p. 49
77. André, J. M.: Adv. Quant. Chem. *12*, 65 (1980)
78. Kertesz, M.: Adv. Quant. Chem. *15*, 161 (1982)
79. Karpfen, A.: Phys. Scripta, *T1*, 79 (1982)
80. Atoji, M., Lipscomb, W. N.: Acta Cryst. *7*, 173 (1954)
81. Johnson, M. W., Sandor, E., Arzi, E.: Acta Cryst. *B31*, 1998 (1975)
82. Kittelberger, J. S., Hornig, D. F.: J. Chem. Phys. *46*, 3099 (1967)
83. Anderson, A., Torrie, B. H., Tse, W. S.: Chem. Phys. Lett. *70*, 300 (1980)
84. Axmann, A., Biem, W., Borsch, P., Hoszfeld, F., Stiller, H.: Faraday Disc. *7*, 69 (1969)
85. Habuda, S. P., Gagarinsky, Yu. V.: Acta Cryst. *B27*, 1677 (1971)
86. Karpfen, A.: Chem. Phys. *47*, 401 (1980)
87. Beyer, A., Karpfen, A.: Chem. Phys. *64*, 343 (1982)
88. Sandor, E., Farrow, R. F. C.: Nature, *215*, 1265 (1967)
89. Diercksen, G. F., Kraemer, W. P., Roos, B. O.: Theoret. Chim. Acta *36*, 249 (1975)
90. Curtiss, L. A., Frurip, D. J., Blander, M.: J. Chem. Phys. *71*, 498 (1979)
91. Chupka, W. A., Russel, M. E.: J. Chem. Phys. *49*, 5426 (1967)
92. Weise, H. P.: Ber. Bunsenges. Phys. Chem. *77*, 578 (1973)
93. Böhme, D. K., in: Interactions between Ions and Molecules (ed.) Ausloos, P., p. 723, New York, Plenum Press 1975
94. Wolf, J. F., Staley, R. H., Koppel, I., Taagepera, R. T., McIver, jr., R. T., Beauchamp, J. L., Taft, R. W.: J. Am. Chem. Soc. *99*, 5417 (1977)
95. Yamdagni, R., Kebarle, P.: J. Am. Chem. Soc. *98*, 1320 (1976)
96. Foster, M. S., Beauchamp, J. L.: Inorg. Chem. *14*, 1229 (1975)
97. Polley, C., Munson, M. S. B.: Abstr. 24th Conference on Mass Spectrometry, p. 723, San Diego, Calif., May 1976
98. Hogg, A. M., Kebarle, P.: J. Chem. Phys. *43*, 449 (1965)

99. Herzberg, G.: Spectra of diatomic molecules, Molecular Spectra and Molecular Structure (ed.) Vol. II, Van Nostrand 1950
100. Hopkinson, A. C., Holbrook, N. K., Yates, K., Csizmadia, I. G.: J. Chem. Phys. 49, 3596 (1968)
101. Alagona, G., Scrocco, E., Tomasi, J. Theoret. Chim. Acta 47, 133 (1978)
102. Szczesniak, M., Ratajczak, M.: Chem. Phys. Lett. 74, 243 (1980)
103. Karpfen, A.: in preparation
104. Diercksen, G. H. F., Kraemer, W. P.: Chem. Phys. Lett. 6, 419 (1970)
105. Lischka, H.: J. Am. Chem. Soc. 96, 4716 (1974)
106. Kollman, P., McKelvey, J., Johansson, A., Rothenberg, S.: J. Am. Chem. Soc. 97, 955 (1975)
107. Clementi, E.: J. Chem. Phys. 46, 3851 (1967)
108. Raffenetti, R. C., Phillips, D. H.: J. Chem. Phys. 71, 4534 (1979)
109. Smith, D. F.: J. Mol. Spectry. 3, 473 (1959)
110. Benedict, W. S., Gailar, N., Plyler, E. K.: J. Chem. Phys. 24, 1139 (1956)
111. Tursi, A. J., Nixon, E. R.: J. Chem. Phys. 52, 1521 (1970)
112. Huber, K. P., Herzberg, G.: Constants of diatomic molecules, Molecular Spectra and Molecular Structure, (ed.), Vol. 4, New York, Van Nostrand, Reinhold Publ. Co., 1979
113. Daudey, J. P., Novaro, O., Kolos, W., Berrondo, M.: J. Chem. Phys. 71, 4297 (1979)
114. Dobbs, R. E., Jones, G. O.: Reps. Prog. Phys. 20, 560 (1957)
115. Horton, G. K.: in: Rare Gas Solids (eds.) Klein, M. L., Venables, J. A., Vol. 1, p. 87, New York, Academic Press 1976
116. Slater, J. C.: Quantum Theory of Molecules and Solids, Vol. 2, p. 334, New York, McGraw Hill 1965
117. Eisenberg, D., Kauzmann, W.: The Structure and Properties of Water, Oxford, Clarendon Press 1969
118. Savoie, R., Anderson, A.: J. Chem. Phys. 44, 548 (1966)
119. Sandorfy, C.: Topics in Current Chemistry, 120, p. 41 (1984)

Vibrational Spectra of Hydrogen Bonded Systems in the Gas Phase

Camille Sandorfy

Département de Chimie, Université de Montréal, Montréal, Québec, Canada, H3C 3V1

Table of Contents

Camille Sandorfy

1 Introduction

Theoreticians did by far the most gas phase work on hydrogen bonding; the results of the great majority of quantum chemical calculations on H-bonds apply to isolated systems. (See [1] as a general reference on H-bonding.) It was unfortunate but inevitable that they were usually compared to spectra obtained in solution or in the liquid or solid phase. While this does not necessarily invalidate the conclusions that can be drawn from such comparisons between theory and experiment it is known that the properties of H-bond systems are modified by perturbations due to additional intermolecular interactions with solvent molecules [2] or with their environment in a liquid or in a crystal.

The aim of the present chapter is to review our knowledge on H-bonds in the gas phase as obtained from vibrational spectroscopy, mainly infrared. Only occasional references will be made to quantum chemical calculations.[1]

Many of the characteristic properties of H-bonds can be deduced from vibrational spectra in which bands due to the "free", non-hydrogen bonded species are, in most cases, readily distinguished from those due to the associated, H-bonded species.

Let at first consider the simple linear triatomic systems

$$X - H \ldots Y$$

where X is a proton donor and Y is a proton acceptor. This unit has $3N-5 = 4$ degrees of vibrational freedom. Various nomenclatures are used in the literature; we shall adopt the following one:

X-H stretch ν_1 the high frequency OH, NH, ... stretch
X-H ... Y bend ν_2 (δ, γ) bending of the X-H bond out of the XHY line
XH ... Y stretch ν_3 or ν_σ the bridge stretching motion. ν_1 and ν_2 are essentially internal vibrations of the proton donor. The bending ν_2 is a degenerate vibration for the linear triatomic unit. Under lower symmetry it will split into an in-plane and an out-of-plane bending motion. If X and Y represent actual molecules there are of course several other vibrations.

Next we consider a H-bonded complex in which the proton donor is diatomic, like $(CH_3)_2O.HCl$. Dimethylether, a non-linear molecule has 6 degrees of rotational and translational freedom while HCl, a linear molecule has 5. The complex which is non-linear will have 6 such degrees of freedom. Thus when the complex is formed $(6 + 5 - 6) = 5$ degrees will be transformed into vibrational degrees of freedom. These intermolecular vibrations are shown in Fig. 1. One of these is ν_σ (or ν_3) which in most known cases is between 100 and 200 cm^{-1} for weak or moderately strong H-bonds. In addition there will be two degenerate bridge deformation vibrations. One of these, ν_b is in the 500–700 cm^{-1} region in the cases of interest

---

1 For a review of a different conception the Reader may consult the chapter by Y. Maréchal [101] which appeared since the manuscript of the present one has been completed.

to us, the other, ν_β, below $100\,\mathrm{cm}^{-1}$. Under less than C_3 symetry the degenerate bands will be split.

If none of the partners is linear 6 degrees of rotational and translational degrees of freedom will be transformed into vibrational ones. The additional intermolecular vibration is a torsional motion. If the proton donor is polyatomic there will be ν_1 and ν_2 and other essentially intramolecular vibrations while in the case of HCl only ν_1 can exist.

These are the vibrations most affected by H-bonding; we shall often encounter them in this review.

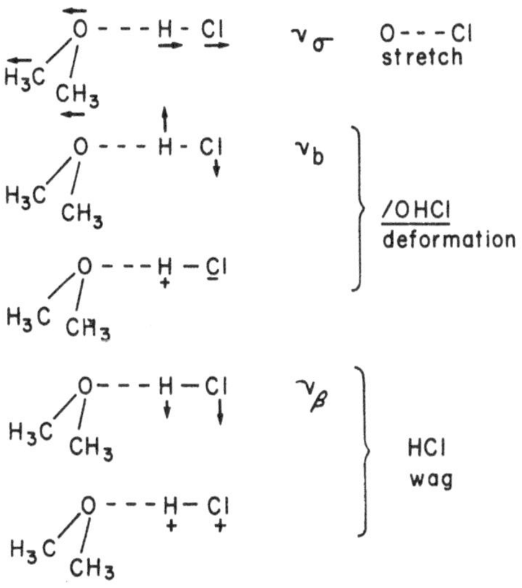

Fig. 1. The Intermolecular (H-bridge) vibrations in the $(CH_3)_2O.HCl$ complex. From J. E. Bertie and M. V. Falk. Can. J. Chem. *51*, 1713 (1973). Reproduced by permission from the National Research Council of Canada

A spectral band is characterised by its frequency range, its intensity, and its shape and breadth. The frequency change in the XH stretching band ($\Delta\nu$) is often used as an approximate measure of the strength of the H-bond. Enthalpies of H-bond formation are usually determined from the temperature variation of the free-associated intensity ratio of the same bands. Of even greater interest are, however, the geometries and the potential surfaces of H-bonded species and the distribution of the electronic charge therein. For this the spectra of the isolated H-bond complexes that is, gas phase spectra are needed. The fine structure of the bands has to be examined. Key questions are: how do the new vibrational motions introduced by the formation of a H-bond interact with the internal motions of the components X and Y? How could this be inferred from the observed breadth and fine structure of the bands?

Most theories of H-bonding have as their starting point the interaction of the "fast" X-H stretching motion (ν_1) and the „slow" XH ... Y bridge stretching motion (ν_3 or ν_σ). This is a useful working hypothesis but a somewhat simplified view. As already stated there are other low frequency H-bond vibrations with which ν_1 can interact.

The X-H stretching vibration is by far the most studied among the vibrations of H-bonded complexes. Its fundamental is distinguished by its high intensity and great breadth compared to the free band.

Anharmonicity manifests itself in a variety of ways around this band. In many cases v_1 is accompanied by a number of overtones and combination tones of vibrations of lower frequency whose intensity is boosted by Fermi resonance with v_1. Ammonium and amine salts, [4-7] carboxylic acid dimers [8] are well known examples of this. In most cases these bands can only be seen with H-bonds of more than average strength, not with alcohols, amines, for example. As will be discussed later, in the gas phase the breadt of the main band (v_1) is due to combinations between v_1 and the low frequency H-bond motions mentioned above and hot bands of these. The existence of combination bands always requires anharmonic coupling as does Fermi-resonance and resonances of higher degree.

The number of H-bonded complexes whose spectra have been determined in the gas phase is relatively limited. This has natural reasons. The complex should be sufficiently volatile; moderate heating may help but it may lead to dissociation. At too high a pressure a variety of associated species might be present; pressure broadening may obscure fine structure. Strong H-bonds involving ions are difficult to put into the gas phase; they form solid complexes or salts with low vapor pressures and decompose at higher temperatures. Very weak H-bonds simply dissociate in the gas phase. Thus all known examples of well studied H-bonds in the gas phase are moderately strong but not very weak or very strong. These limitations make us choose to follow the chronological order of development in this field. Three stages can be distinguished: in the first period vibrational analysis was applied; in the second one arguments stemming from rotational fine structure of the vibrational bands were added; in the third period a wealth of information from pure rotational spectra became available.

2 Preliminaries

2.1 Mechanical Anharmonicity

Vibrational problems are usually treated within the framework of the harmonic oscillator approximation. It implies a quadratic potential function, wavefunctions whose characteristic parts are the exact Hermite polynomials, zero vibrational amplitudes, and the mutual independence of the 3N-6 (or 3N-5) normal vibrations. Whereas no molecule is exactly a harmonic oscillator, this approximation has been highly successful: most of our present knowledge on molecular vibrations has been gained through applying it to actual problems. There are cases, however, which require higher approximations even for a qualitative understanding of conditions. Hydrogen bonding is quite certainly one of these. To convince oneself it is sufficient to look at the usual triatomic model:

$$X - H \dots Y .$$

Since the proton motion is influenced by two centers of attraction, it would be hard to understand how it could have a harmonic potential.

This does not mean that the whole concept of normal vibrations must be dispensed with. There is evidence [9,10] that for weak H-bonds anharmonicity is just about the same as for the related "free" vibrations. For medium strong H-bonds like those in multimeric alcohols and up to carboxylic acid dimers second order perturbation theory is adequate for at least the lower vibrational levels. Basic knowledge for this is found in Herzberg [11]. A review on anharmonicity problems in connection with H-bonds has been given by the writer [12]. All available evidence from gas and solution spectra shows that H-bond formation increases the anharmonicity of the X-H stretching vibration and introduces appreciable anharmonic coupling between the X-H and XH ... Y stretching motions. Furthermore there is at least qualitative evidence that the stronger the H-bond the larger are the anharmonic constants [12].

To the second order, for a diatomic oscillator, the potential governing the vibrational motion can be expressed as

$$V = \frac{1}{2} kQ^2 + k_3Q^3 + k_4Q^4 \tag{1}$$

where the first term is the harmonic potential and the higher terms represent (mechanical) anharmonicity. Q is the normal coordinate and k, k_3 and k_4 are the harmonic, cubic and quartic potential constants respectively. For the vibrational terms (G_v), second order perturbation treatment yields

$$G_v = \omega_e \left(v + \frac{1}{2} \right) + X \left(v + \frac{1}{2} \right)^2 \tag{2}$$

where ω_e is the harmonic frequency (which would be the frequency if the oscillator was perfectly harmonic) v is the vibrational quantum number and X is the anharmonicity constant. G_v, ω_e and X are expressed in wavenumber units, cm^{-1}. The fundamental frequency is $v^{01} = G_1 - G_0$, the first overtone is $v^{02} = G_2 - G_0$, the second overtone is $v^{03} = G_3 - G_0$ and so on:

$$\begin{aligned} v^{01} &= \omega_e + 2X \\ v^{02} &= \omega_e + 6X \\ v^{03} &= \omega_e + 12X \, . \end{aligned} \tag{3}$$

From these

$$X = \frac{1}{2} v^{02} - v^{01} = \frac{1}{3} v^{03} - v^{01} \tag{4}$$

and it is given the negative sign if, as is normally the case, anharmonicity decreases the frequency. X can be computed if the frequencies of the fundamental and an overtone are known. In this simple case X is related to the potential constants through

$$X = 15k_3^2/4\omega_e - 3/2k_4 \, . \tag{5}$$

Important new features are introduced when polyatomic oscillators are considered. Let us take the simple triatomic model X-H ... Y and call v_1 and v_3 the X-H and

Camille Sandorfy

XH ... Y stretching vibrations respectively, v_2 the bending vibration and disregard the degeneracy of the latter. Then we obtain for a vibrational term, to the second order:

$$
\begin{aligned}
G(v_1, v_2, v_3) = & \; \omega_1 \left(v_1 + \frac{1}{2} \right) + \omega_2 \left(v_2 + \frac{1}{2} \right) + \omega_3 \left(v_3 + \frac{1}{2} \right) \\
& + X_{11} \left(v_1 + \frac{1}{2} \right)^2 + X_{22} \left(v_2 + \frac{1}{2} \right)^2 + X_{33} \left(v_3 + \frac{1}{2} \right)^2 \\
& + X_{12} \left(v_1 + \frac{1}{2} \right) \left(v_2 + \frac{1}{2} \right) + X_{13} \left(v_1 + \frac{1}{2} \right) \left(v_3 + \frac{1}{2} \right) \\
& + X_{23} \left(v_2 + \frac{1}{2} \right) \left(v_3 + \frac{1}{2} \right)
\end{aligned}
\tag{6}
$$

It contains six anharmonicity constants. The three having two identical indices belong to one of each of the three normal vibrations, the three others are the coupling constants which give a measure of their interdependence. The frequency in (cm^{-1}) of a given vibrational transition is:

$$
v = G'(v_1', v_2', v_3') - G''/v_1'', v_2'', v_3'')
\tag{7}
$$

where for bands other than hot bands, $v_1'' = v_2'' = v_3'' = 0$. For the fundamental of v_1 we obtain:

$$
v_1^{01} = \omega_1 + 2X_{11} + \frac{1}{2} X_{12} + \frac{1}{2} X_{13}
\tag{8}
$$

For a diatomic oscillator only the first two terms would appear but now we see that the coupling constants between Q_1 and the other two vibrations also have a bearing on the frequency. The first overtone of the same vibration is:

$$
v_1^{02} = 2\omega_1 + 6X_{11} + X_{12} + X_{13} = 2\omega_1^{01} + 2X_{11}
\tag{9}
$$

so that if we measure the overtone and the fundamental we can compute X_{11} just as for a diatomic oscillator.

Because of anharmonic coupling two or more vibrations may be excited by absorption of just one photon. These combination bands are essential for the understanding of H-bond systems. If $(v_1 + v_3)$ denotes the binary combination of the fast $v_1(X - H)$ and slow $v_3(XH ... Y)$ stretching vibrations we have:

$$
\begin{aligned}
& (v_1 + v_3) = G'(1, 0, 1) - G''(0, 0, 0), \\
& (v_1 + v_3) = \omega_1 + \omega_3 + 2X_{11} + 2X_{33} + \\
& 2X_{13} + \frac{1}{2} X_{12} + \frac{1}{2} X_{23} = v_1^{01} + v_3^{01} + X_{13}
\end{aligned}
\tag{10}
$$

Thus if we measure the frequency of the combination band and the two fundamentals we can compute the coupling constant X_{13}.

This is a summation tone. The difference tone of v_1 and v_3 is obtained if v_3 is at level $v_3 = 1$ when the photon strikes. (Hot band).

$$(v_1 - v_3) = G'(1, 0, 0) - G''(0, 0, 1),$$

$$(v_1 - v_3) = \omega_1 - \omega_3 + 2X_{11} - 2X_{33} + \frac{1}{2}X_{12} - \frac{1}{2}X_{23}$$

$$= v_1^{01} - v_3^{01} \tag{11}$$

Interestingly, the coupling constant cancels out and the wavenumber of the difference band is simply the difference of the wavenumbers of the two fundamentals.

A very important type of combination band is obtained when a transition, say $0 \to 1$ of a vibration of high frequency (like v_1) combines with $1 \to 1$, $2 \to 2$, $3 \to 3$, etc. transitions of a vibration of low frequency (like v_3 or the bridge deformation modes.) Then a series of bands is obtained which is analogous to "sequences" known from electronic spectroscopy. These are denoted as $(v_1 + n'v_3 - n''v_3)$ or, in this case,

$$(v_1 + v_3'v_3 - v_3''v_3). \tag{12}$$

With $v_3' = 1$ and $v_3'' = 1$ one obtains

$$(v_1 + v_3 - v_3) = G'(1, 0, 1) - G''(0, 0, 1)$$

$$= \omega_1 + 2X_{11} + \frac{1}{2}X_{12} + \frac{3}{2}X_{13} = v_1^{01} + X_{13} \tag{13}$$

So the separation from the v_1^{01} fundamental is just X_{13}, the coupling constant! These bands are, of course, hot bands, their intensity depending on the Boltzmann factor. Since X_{13} can be either positive or negative such hot bands may appear on either the low or the high frequency side of v_1^{01}. Furthermore, since the coupling constant X_{13} is usually relatively small such bands can make a considerable contribution to the apparent width and intensity of the main v_1^{01} band. (For additional examples see [12] and [13].)

As will be elaborated on later these combination bands are the key to the understanding of the infrared spectra of H-bond systems.

2.2 Electrical Anharmonicity

Overtones and combination tones are forbidden in the harmonic oscillator approximation. Mechanical anharmonicity is one factor that might give them intensity through violating the $\Delta v = \pm 1$ selection rule. There is another possible cause for this, however, which might operate even when the oscillator is perfectly harmonic. It is electrical anharmonicity.

The transition moment R can be written in the form:

$$R = \int \Psi_{v'} \mu \Psi_{v''} \, d\tau. \tag{14}$$

The (instantaneous) dipole moment μ varies during the vibration. It can be expanded into a Taylor series at the equilibrium geometry:

$$\mu = \mu_e + (\partial\mu/\partial Q)_e\, Q + (\partial^2\mu/\partial Q^2)_e\, Q^2 + (\partial^3\mu/\partial Q^3)_e\, Q^3 + \ldots \qquad (15)$$

where the sum of the higher than linear terms is called the electrical anharmonicity. With these the transition moment becomes:

$$R = (\partial\mu/\partial Q)_e \int \Psi_{v'} Q \Psi_{v''}\, d\tau + (\partial^2\mu/Q^2)_e \int \Psi_{v'} Q^2 \Psi_{v''}\, d\tau$$
$$+ (\partial^3\mu/\partial Q^3)_e \int \Psi_{v'} Q^3 \Psi_{v''}\, d\tau + \ldots \; . \qquad (16)$$

In the polyatomic case many terms come in which contain derivatives with respect to two or more normal coordinates.

Thus electrical anharmonicity is the non-linear part of the variation of the dipole moment with normal coordinates. There is no reason to expect it to be negligibly small. It can give intensity to overtones and combination tones. In the general case both mechanical and electrical anharmonicities contribute to the intensity. As will be mentioned later they cannot be disregarded when spectra of H-bond complexes are dealt with.

2.3 Stepanov Diagrams and the Sheppard Effect

The combinations between the X-H and XH ... Y stretching vibrations (v_1 and v_3) are often represented by Stepanov's diagrams [14, 15]. These are similar to the familiar Franck-Condon schemes of electronic spectroscopy. The $\dot{v} = 0$ and $v = 1$ potential curves of the high frequency vibration play the role of the potential curves of the electronic states, the energy levels of the combining low frequency vibration being drawn into these as straight lines (Fig. 2). Sheppard pointed out in 1958 [16] that because of the anharmonicity of the X-H stretching vibration the proton will be, on the average, nearer to Y in the $v_1 = 1$ state than in the $v_1 = 0$ state and, as a consequence, the H-bond will be stronger in the $v_1 = 1$ state. Thus the upper potential energy curve is expected to have a deeper minimum than the lower one, and its minimum occurs at a lower value of the X-Y distance. This is a generally accepted principle.

Then one way of regarding the IR spectrum is to say that v_3 gives a fine structure to v_1. Now, because of the Sheppard effect and the Franck-Condon vertical the $(0, 0)\rightarrow(1, 0)$ band may not be the most intense line. (In each bracket the first number represents the vibrational quantum number of v_1 and the second that of v_3.) This implies that non-vertical transitions might have quite appreciable probabilities.

It is good to remember too that we are dealing with combination tones. For example $(0, 0)\rightarrow(1, 1)$ or $(v_1 + v_3)$ is a binary combination (summation) band. Should $(0, 0)\rightarrow(1, 1)$ be more intense than $(0, 0)\rightarrow(1, 0)$ that means that the combination is more intense than the fundamental. This is quite possible but it requires a large X_{13} coupling constant or/and a large amount of electrical anharmonicity. Finally, whether or not $(0, 0)\rightarrow(1, 0)$ is the strongest band among the subbands due to combinations of v_1 and v_3 with various quantum numbers depends on several factors. It is important too that, as Robertson pointed out recently [17], conditions are different for deuterium bonds. Because of the lesser amplitude and mechanic anharmonicity the Sheppard effect is expected to be much smaller. This would tend to make the $(0, 0) \to (1, 0)$ band the most intense one even if this is not so for the related H-bond complex. It is, indeed, a general observation that the ν_1 band has different fine structure for related H-bond and D-bond systems. (See Sects. 3 and 4.)

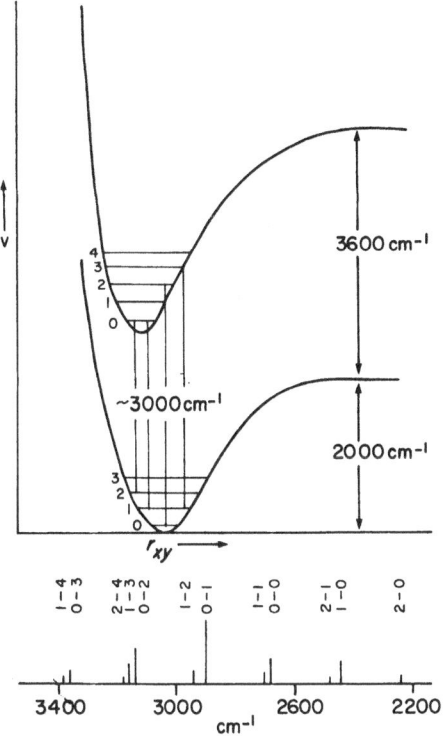

Fig. 2. The Stepanov energy level scheme for the interaction of vXH and vXH ... Y vibrations and a schematic spectrum showing frequencies of the type vXH $\pm$ nvXH ... Y. The lower curve of potential energy (V) v. X ... Y distance (r_{XY}) corresponds to the "ground" state of the vHX vibration, the upper curve to the excited state reached by the absorption of a quantum of energy equal to hvXH. The energy scale is drawn approximately correctly for an OH ... O system of moderate strength. vXH free = 3600 cm^{-1}; vXH ... Y = 200 cm^{-1}; vXH (H-bonded) $\approx$ 3000 cm^{-1}; dissociation energy of H-bond in ground state = 6 kcal. per mole ($\approx$ 2000 cm^{-1}). This figure and its legend are taken from a chapter by N. Sheppard, in Hydrogen Bonding. Ed. D. Hadži. 1959. Reproduced by permission from Pergamon Press, London

2.4 Molecular Rotation

A few concepts concerning molecular rotation will be needed. In this respect we refer to Herzberg in a general way [11,18] (Vol. I, chapters III 1 and III 2, Vol. II., chapter IV 1.) A rotational term F(J) is, for a diatomic molecule considered as a rigid rotator,

$$F(J) = \frac{h}{8\pi^2 cI} J(J + 1) = BJ(J + 1) \tag{17}$$

where J is the rotational quantum number, I the moment of inertia and B the rotational constant:

$$B = \frac{h}{8\pi^2 cI} \tag{18}$$

From (17) for the frequency of a rotation line we have:

$$v = F(J') - F(J'') = BJ'(J + 1) - BJ'(J' + 1) \tag{19}$$

or, taking into account the selection rule $\Delta J = \pm 1$ and writing $J = J''$ (for absorption),

$$v = 2B(J + 1); \quad J = 0, 1, 2, \dots . \tag{20}$$

Thus it is seen from [17] that the energy levels of the rigid rotator have an approximate quadratic dependence on J and from [20] that, in spite of this, the rotational spectrum consists of a series of equidistant lines. For a non-rigid rotator a small correction term has to be introduced because of the influence of the centrifugal force, but this will not concern us here.

For a vibrating rotator the rotational constant B has to be replaced by a mean value,

$$B_v = \frac{h}{8\pi^2 c\mu} \overline{\left[\frac{1}{r^2}\right]} \tag{21}$$

where $\overline{\left[\dfrac{1}{r^2}\right]}$ is the mean value of $\dfrac{1}{r^2}$ during the vibration and μ is the reduced mass. In most cases the average value of r is larger than its equilibrium value r_e, because of the anharmonicity of the vibration. Furthermore it is larger for $v = 1$ than for $v = 0$. The frequency of a rotational line in the fine structure of a vibrational band is given by

$$v = v_0 + B_v J'(J' + 1) - B_{v''} J''(J'' + 1) \tag{22}$$

where v_0 is the frequency of the purely vibrational transition.

The selection rule depends on the electronic state; if Λ, the quantum number for the projection of the electronic orbital angular momentum is 0 the rule is $\Delta J = \pm 1$, for $\Lambda \neq 1$ it is $\Delta J = 0, \pm 1$.

If $\Lambda = 0$, the fine structure will contain two branches of rotational lines on both sides of the missing ν_0 called the R and P branches for $\Delta J = +1$ and $\Delta J = -1$.

$$\nu_R = \nu_0 + 2B'_v + (3B'_v - B''_v) + (B'_v - B''_v) J^2 ; \quad J = 0, 1, 2, ... \quad (23)$$
$$\nu_P = \nu_0 - (B'_v + B''_v) J + (B'_v - B''_v) J^2 ; \quad J = 1, 2, ... \quad (24)$$

Since $B'_v < B''_v$ and since as J increases the quadratic term overtakes the linear term [23] (they are inversely proportional to r^2) there is a bandhead in the R branch that is at the high frequency side of ν_0. For $\Lambda \neq 0$ there is a third, the Q branch for $\Delta J = 0$ which is absent or weak for parallel transitions ($\Delta \Lambda = 0$) and strong for perpendicular transitions. ($\Delta \Lambda = \pm 1$).

We are not going into any further detail. It is important to point out, however, that for H-bonded systems the Sheppard effect makes the X ... Y distance shorter (not longer) for $v = 1$ than for $v = 0$ and for this reason $B'_v > B''_v$. Thus we expect to find the bandhead not in the R but in the P branch. As will be seen this expectation is born out by experiment.

3 Hydrogen Bonding in Gaseous Mixtures

It all started in 1965 with a series of six papers by Millen and his coworkers at University College, London [19-25]. (Although they had a preliminary communication in 1961.) This has been a conscient effort to reach the isolated H-bond itself, free of environmental effects due to solvent, or liquid or crystalline neighborhood. "These systems provide an opportunity of obtaining a better understanding of the qualitative features of the spectra of hydrogen bonded complexes and a possible route to the development of a quantitative understanding of certain aspects, particularly spectral intensities, molecular structure and potential functions". (Bertie and Millen [19]). They also recognised the key importance of the exceptional breadth of infrared bands (mainly ν_1) affected by H-bond formation. It indeed contains most of the secrets about H-bonds. The mixtures they studied were the following:

        ether-hydrogen chloride (Bertie and Millen [19])
        ether-hydrogen fluoride (Arnold and Millen [20])
        carbonyl compounds-hydrogen fluoride (Arnold and Millen [21])
        ether-deuterium chloride (Bertie and Millen [22])
        ether-nitric acid systems (Millen and Samsonov [23])
        amine-alcohol systems (Millen and Zabicky [24]).

Later developments made the two first systems especially important and we are going to describe their spectra at this point.

3.1 The Dimethyl Ether-Hydrogen Chloride Complex

The well known spectrum obtained with 250 Torr of HCl and 125 Torr of $(CH_3)_2O$ with a 10 cm cell taken from [19] is shown in Fig. 3. Under the given experimental conditions the partial pressures that had to be used are not as low as would be desirable in view of the highly associable nature of the system and we cannot quite consider the mixture as an ideal gas. Some of the observed breadth of the bands is likely to be due to pressure broadening. However, the spectrum reveals the existance of four fairly broad bands. By varying the relative partial pressures Bertie and Millen were able to show that the spectrum is due to a 1:1 complex. The observed peaks cannot correspond to partly resolved rotational structure (P, Q. R branches); the band is a parallel band and the spacing is much too large. They are not due either to unassociated ether or hydrogen chloride. Similar results were obtained with several other ethers.

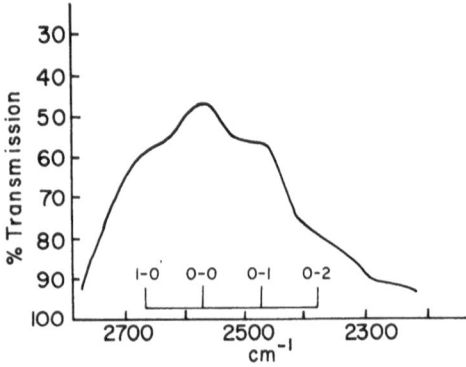

Fig. 3. The infrared spectrum of the dimethylether-hydrogen chloride complex in the gas phase. From J. E. Bertie and D. J. Millen, J. Chem. Soc. 497 (1965). Reproduced by permission from the Chemical Society

The central peak for dimethyl ether has its maximum at 2570 cm^{-1}, the subbands (also called satellites) or shoulders being at about 90–100 cm^{-1} on each side of it. The observed frequencies are about 2660, 2570, 2480 and 2360 cm^{-1}. In their original paper Bertie and Millen [19] assigned the most intense peak at 2570 cm^{-1} to the H-bonded Cl-H ... O stretching vibration. (v_1) Then what are the other peaks? Comparison of the spectra of complexes of HCl with different ethers and alkyl deuterated ethers shows that they are not combination bands of ether vibrations of lower frequency boosted by Fermi resonance. While the frequencies of these bands vary from one ether to the other the pattern shown in Fig. 3 remains practically the same. The only remaining possibility is that they are combinations of the main v_1 band with the low frequency bridge vibrations of (v_σ or v_β) which now appear as a major cause of the broadening of v_1. Thus Bertie and Millen assigned the observed peaks to sum and difference combinations of v_1 and the bridge stretching vibration, v_σ:

$$2660 \text{ cm}^{-1} \quad (0, 0) \rightarrow (1, 1)$$
$$2570 \quad (0, 0) \rightarrow (1, 0)$$
$$2480 \quad (0, 1) \rightarrow (1, 1)$$
$$2360 \quad (0, 2) \rightarrow (1, 2)$$

where in the brackets the first and second number refer to the vibrational quantum numbers of v_1 and v_σ, respectively. Since the apparent half width of these bands is larger than what could be explained through the rotational contours the Authors suggested that hot bands could contribute to the band width. This has been later substantiated through Thomas' work [76-79].

The essential result of this work has been the appearance of subbands around the H-bonded H-Cl stretching band in the gas phase. Their interpretation as $(v_1 \pm nv_\sigma)$ combination bands is still considered essentially correct although other combinations, among them hot bands are also involved.

The corresponding DCl complexes were examined by the same Authors [22]. They found that the v_1 bands extend from about 2000 to about 1600 cm^{-1} and that the side bands are much weaker. This could be explained through a consideration of the effect of mechanical anharmonicity on the intensities of sum and difference bands which is mass dependent. The spacing of the subbands should be otherwise the same for HCl and for DCl since the O ... Cl motion is only slightly affected by the replacement of H by D.

As stated above this spacing is about 90–100 cm^{-1}. This should correspond to the v_σ bridge stretching frequency. It has been measured to be 119 cm^{-1} by Belozerskaya and Shchepkin [26].

The assignments for the dimethylether-hydrogenchloride complex were later modified by Lassègues and Huong [27] and by Bertie and Falk [28]. The former varied temperature from 90° to −50 °C. Since according to the original assignments of Bertie and Millen [19] the 2470 cm^{-1} band is a difference band its intensity would be expected to decrease upon lowering the temperature. It was found, however, that it dit not. Consequently Lassègues and Huong reassigned the 2470 cm^{-1} band as the $(0, 0) \rightarrow (1, 0)$ band and the strongest peak at 2570 cm^{-1} as the $(0, 0) \rightarrow (1, 1)$ band. Bertie and Falk who also varied temperature although through a narrower range came to the same conclusion. These new assignments seem to be at present generally accepted. In the writer's opinion, however, some caution should be exercised. All these bands receive a great deal of intensity from hot bands of either v_σ or the very low frequency bridge deformation frequency, v_β. So the temperature dependence of their intensity does not depend only on the Boltzmann factor of v_σ. Furthermore there is a great deal of mutual overlap between these bands. Suppose, for example that the 2470 cm^{-1} band is a difference band $(0, 1) \rightarrow (1, 0)$. At lower temperatures it would lose intensity. This would cause an increase in the intensity of the $(0, 0) \rightarrow (1, 0)$ band which would then be the one at 2570 cm^{-1}. Because of the mutual overlap between the two bands this would cancel much of the loss of the difference band.

For the time being the balance of evidence makes us adopt the new assignments (2470 cm^{-1} for the 0, 0 band) but not without reservation. They should certainly not be generalized to other systems without further study. Bertie and Falk [28] assigned the two components of the higher degenerate bridge deformation (v_b) to bands found at 525 and 470 cm^{-1}. (See the next section for HF). They also made an important observation on the DCl complex. In the IR spectrum of the latter the $v_1 \pm v_\sigma$ combinations are not evident. As stated above this can be explained by the weakness of these combinations due to the lesser anharmonicity of DCl motions. A number of shoulders are apparent on the band envelop, however. Bertie and Falk made the suggestion that these are due to $v_1 \pm v_\beta$ combinations involving the low frequency bridge deformation vibration whose approximate frequency must be about 50 cm^{-1}. This is very important. If there are such combination bands on the v_1 band of the DCl complex, they could exist for the v_1 band of the HCl complex as well; they cannot be identified with certainty because of overlap from the stronger $v_1 \pm v_\sigma$ bands. The breadth of the latter shows, however, that there might well be other bands under the overall band envelop. As will be mentioned later there are other facts that underscore the role of the combinations involving v_β.

It may seem to be surprising at first sight that some of the combination bands are more intense than the (v_1) fundamental itself. As mentioned in the previous section because of the anharmonicity of v_1 (the XH stretching vibration) in the $v_1 = 1$ state of v_1 the proton will be, on the average, closer to Y than in the $v_1 = 0$ state. A Franck-Condon type reasoning then shows that certain combinations might have considerable intensity and this despite an unfavorable Boltzmann factor. In keeping with this Bertie and Falk assigned the 2660 cm^{-1} band to two overlapping transitions: $(0, 0) \rightarrow (1, 2)$ and $(0, 2) \rightarrow (1, 4)$. The latter is a hot band and it may be partly responsible for the temperature sensitivity of 2660. Lassègues and Huong [27] made similar assignments:

$$2660 \text{ cm}^{-1} \quad (0, 0) \rightarrow (1, 2); (0, 1) \rightarrow (1, 3); (0, 2) \rightarrow (1, 4); \dots$$
$$2570 \text{ cm}^{-1} \quad (0, 0) \rightarrow (1, 1); (0, 1) \rightarrow (1, 2); \dots$$
$$2470 \text{ cm}^{-1} \quad (0, 0) \rightarrow (1, 0); (0, 1) \rightarrow (1, 1); \dots$$
$$2360 \text{ cm}^{-1} \quad (0, 1) \rightarrow (1, 0); (0, 2) \rightarrow (1, 1); \dots$$

These bands are partly "hot" and, of course many other such combinations could be imagined if for the lower frequency we used v_β instead of v_σ. The calculations of Robertson [17] which take account of both the Franck-Condon and Boltzmann factors substantiate these conditions. For the HCl complex at 300 °K the $(0, 1) \rightarrow (1, 0)$ difference band, the $(0, 0) \rightarrow (1, 0)$ fundamental and the $(0, 0) \rightarrow (1, 1)$ summation band which have relative intensities 0.39, 0.80 and 1,00 in this order receive contibutions from (other) hot bands of 0.16, 0.135 and 0.54 respectively. This means that the summation band can be about as sensitive to temperature as the difference band.

Now, and Robertson pointed out, the Franck-Condon factors may be significantly different for the HCl and for the DCl complexes. Because of the lesser

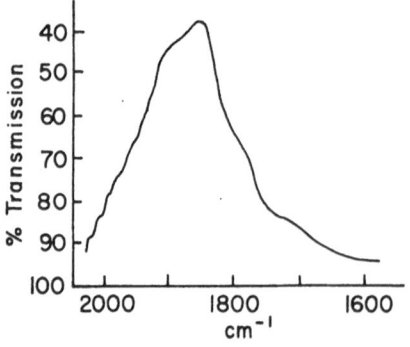

Fig. 4. The infrared spectrum of the dimethylether deuterium chloride complex in the gas phase. From J. E. Bertie and D. J. Millen, J. Chem. Soc. 514 (1965). Reproduced by permission from the Chemical Society

amplitude and mechanical anharmonicity the strengthening of the H-bond at the $v_1 = 1$ level should be much less. This would tend to make the $(0, 0) \rightarrow (1, 0)$ fundamental more intense than the accompanying combination bands. Together with the original explanation of Bertie and Millen this goes a long way in explaining the different appearances of the v_1 regions of the HCl and DCl complexes of dimethylether (Fig. 4).

Another contribution to this problem came from Desbat and Lassègues [29] who succeeded in measuring the *Raman* spectra of both the HCl and DCl complexes in the gas phase at room temperature. At first sight the spectra are similar to the IR spectra: they consist of broad subbands separated by about 100 cm^{-1}. For the DCl complex the subbands are narrower and are more closely spaced, just as in IR (Fig. 5). The Raman and IR profiles are different however. The subbands have their centers at somewhat higher frequencies in Raman and their relative intensities also differ. The most important observation is that for the HCl complex the band at 2470 cm^{-1} which according to the revised assignments is the v_1 fundamental $((0, 0) \rightarrow (1, 0))$ and which is not the most intense among the subbands in the IR, is even less intense in Raman. The relative intensities of the DCl complex are also modified as is seen in Fig. 5.

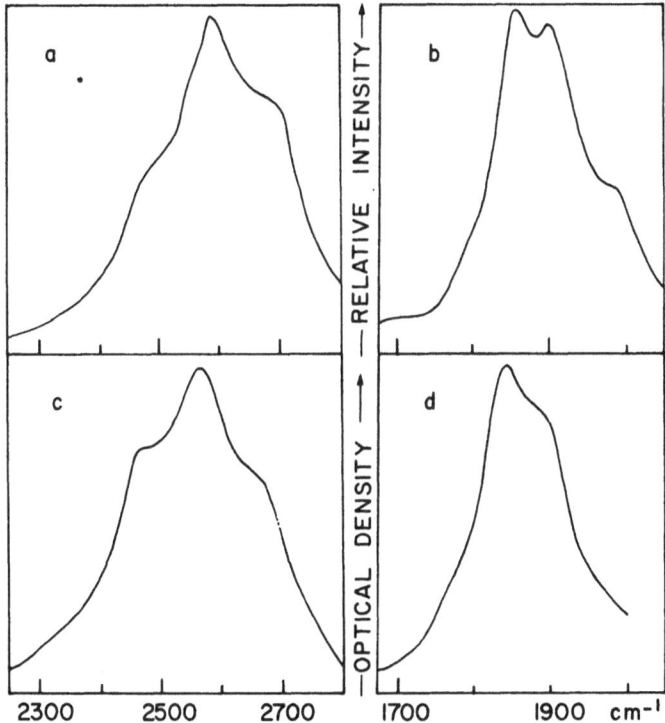

Fig. 5. Comparison of the Ramon (a and b) and infrared (c and d) spectra of the $(CH_3)_2$.HCl (a and c) and $(CH_3)_2$O.DCl (b and d) complexes in the gas phase. From B. Desbat and J. C. Lassègues, J. Chem. Phys. *70*, 1824 (1979). Reproduced by permission from the American Physical Society

What could account for these differences? On the one hand Franck-Condon and Boltzmann factors and mechanical anharmonicity apply to both IR and Raman spectra; on the other hand infrared intensities depend on the rate of change of the dipole moment while Raman intensities depend on the rate of change of polarizability. Furthermore electrical anharmonicity, the non-linear part of the rate of change will also be different for the two kinds of spectra. Electrical anharmonicity has some influence on the intensities of fundamentals and it usually influences to a greater extent the intensities of overtones and combination tones. Unfortunately little is known about the coupling terms in the expansion of the electrical anharmonicity (see Sect. 2.2.) which depend on two or more normal coordinates. The different dependence of IR and Raman spectra on changes in polarizability and dipole moment appears to be a plausible cause of differences in band shape and relative intensities between the two spectra. This also implies electrical anharmonicity which for IR consists of higher derivatives of the dipole moment and, for Raman spectra, of polarizability. Lautié and Novak [30] have insisted on the fact that the XH stretching (v_1) undergoes a much greater enhancement of intensity in IR than in Raman. Bernard-Houplain and Sandorfy [31] have shown that the free: associated intensity ratio is very different in infrared and in Raman: in IR, the associated band is by far the more intense but in Raman the ratio is much more equitable. The free band has a good chance to appear even at concentrations where in the IR it is not or hardly apparent. They found the Raman profile more alike the one of the first infrared *overtone*. As to the latter Di Paolo, Bourdéron and Sandorfy [32] have shown that mechanical and electrical anharmonicities may, under certain conditions cancel each others effects causing the notorious weakness of H-bonded XH overtones. (Luck and Ditter [33].) More recently, Maréchal and Bratos [34] put forward the idea that electrical anharmonicity might be the cause of the differences in profile between H-bonded infrared and Raman bands. It is then conceivable that in Raman the combination bands receive additional intensity by polarizability related electrical anharmonicity while in the IR they receive less from dipole related electrical anharmonicity. The fundamentals themselves may receive different contributions of this kind. It is also possible that in the HCl complex the $v_1 \pm v_\sigma$ combinations are favored whereas in the DCl complex the $v_1 \pm v_\beta$ combinations are. While there is no direct proof for these assumptions they amount to a coherent qualitative picture and are the only ones which are available at the present time.

Extensive theoretical works are now available which substantiate and justify the above assumptions. Robertson [17] based his theory on coupling between the two stretching modes v_1 and v_σ using a two parameter Morse potential. He took account of Franck-Condon factors, the Boltzmann distribution and mechanical but not electrical anharmonicity. He constructed effective potential curves governing the H-bond vibration in the ground and first excited states of v_1. His calculations demonstrate the shortening of the O ... Cl distance in the excited state for the HCl complex of dimethylether and a much less pronounced shortening for the DCl complex. He was also able to account for the formation of $v_1 \pm v_\sigma$ combination bands and the observed temperature effect on the basis of the assignments of Lassègues and Huong [27] and Bertie and Falk [28]. Bouteiller and Maréchal [35] represented the complexes with a linear triatomic model where the v_1 and v_σ vibrations are coupled by an anharmonic potential. The coupling term was determined from the temperature dependence of the spectra. They successfully reproduced the main features of the observed IR spectra. Bouteiller and Guissani [36] used the Witkowski-Maréchal potential [37] to interpret the noticeable decrease in the separation between the ν_1 fundamental and the subbands on deuteration and the difference between the IR and Raman profiles. Involving both mechanical and electrical anharmonicities for the first time they were able to interpret both effects. In particular they found that electrical anharmonicity is responsible for the differences between the observed IR and Raman band shapes.

3.2 The Dimethylether-Hydrogen Fluoride Complex

Just as the dimethylether-hydrogen chloride complex, this system yielded much useful information. Investigation of it started with Arnold and Millen's paper in 1965 [20].

HF forms stronger H-bonds than HCl and F is lighter than Cl. Thus we expect to find the associated HF stretching band at higher frequencies, a larger shift ($\Delta \nu_1$) from the position of the free band and also a higher ν_σ frequency than for the dimethylether-hydrogen chloride complex. All this is born out by experiment. The main associated HF band is found at 3470 cm^{-1} with a weaker subband at 3710 and a still weaker band at about 3300 cm^{-1} (Fig. 6). This spectrum was obtained with 20 Torr of HF, 80 Torr of dimethylether and a 10 cm path length. Pressure variation has showed that the spectrum belongs to the 1:1 complex. By the presence of subbands and their relative breadth this spectral region is similar to the ν_1 region of the HCl complex. The spacing between the three components is, as expected, about 180 cm^{-1}. A band of this frequency was later found by Thomas [38] and is naturally assigned to ν_σ. The side bands in this case are weaker than for the HCl complex. Shoulders are apparent on the band envelops revealing the existence of additional combination bands. Arnold and Millen [20] assigned the strong peak at 3470 cm^{-1} to the ν_1 fundamental ((0, 0)→(1, 0)) and the side-bands at 3710 and 3300 cm^{-1} to the $\nu_1 \pm \nu_\sigma$ combinations, (0, 0)→(1, 1) and (0, 1)→(1, 0), respectively. The assignments of the shoulders, also made by Arnold and Millen are shown in Fig. 6. The appearence of the spectrum similar to the HCl complex, and the larger spacing clearly indicate that the HF spectrum has to be interpreted in the same way as the HCl spectrum. The subbands are weaker, however. With $(CD_3)_2O$ an almost

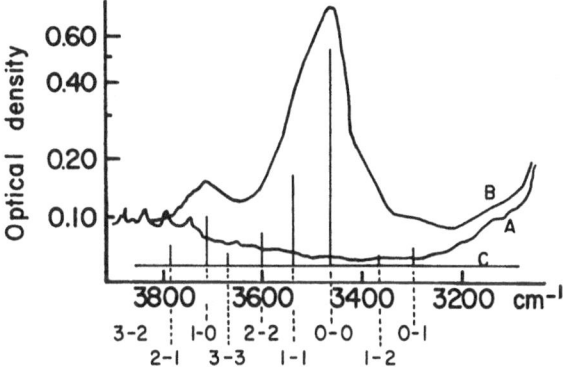

Fig. 6. The infrared spectrum of the dimethylether-hydrogen fluoride complex in the gas phase. From J. Arnold and D. J. Millen, J. Chem. Soc. 503 (1965). Reproduced by permission from the Chemical Society.

identical spectrum was obtained which excludes the possibility that the subbands are due to combinations of ethereal modes boosted by Fermi resonance with v_1.

In the spectrum of the DF complex (with fully deuterated dimethylether) the main band which is clearly not a single band is centered at 2550 cm^{-1}. A weaker band at higher frequencies is separated by about 220 cm^{-1} from the main band what is not so different from that found with the HF complex. This is in line with its O ... F character. Just as in the case of the HCl complex we observe the relative weakness of the subbands of $(CH_3)_2O.DF$ with respect to those of $(CH_3)_2O.HF$ making visible some combinations with a band of even lower frequency which can only be v_β, the lower bridge deformation mode.

The relative weakness of the subbands is very likely connected with the frequencies of v_σ and v_β which are, of course, higher than for the HCl complex. This diminishes the intensity of the hot bands due to the higher Boltzmann factor. Arnold and Millen varied temperature from room temperature to 70 °C and found that the intensity of the 3300 cm^{-1} band increased with increasing temperature more than that of the other bands confirming its assignment as a difference tone. It would be desirable to measure this spectrum at lower temperatures as well but, to the writer's knowledge, this has not been done.

The above discussion implies that the v_1 fundamental, $(0, 0) \rightarrow (1, 0)$, is assigned to the strongest peak at 3470 cm^{-1}. Contrary to the HCl complex this has not been challenged leaving one with the uncomfortable feeling that this matter may still not be settled. Otherwise the general picture emerging from the study of the HF complex is essentially the same as that obtained from the consideration of the vibrational spectrum of the HCl complex. Arnold, Bertie and Millen [19,20,22] examined several other ethers H-bonded with HCl and HF with similar results.

One more comment should be made at this point. It concerns the anharmonic coupling constant between the v_1 and v_σ modes. In the case of the HCl complex the great diffuseness of the 2670 cm^{-1} subband and the argument about the assignments made its evaluation unsafe. For the HF complex the situation is somewhat more favorable. As has been said in Section 2.1 the separation between v_1 and the difference band $(v_1 - v_\sigma)$ is just $v_1 - v_\sigma$ but the separation between v_1 and the summation band $(v_1 - v_\sigma)$ is equal to $v_1 + v_\sigma + X_{1\sigma}$ where $X_{1\sigma}$ is the coupling constant. On the basis of the above assignments its value is

$$X_{1\sigma} = (3710 - 3470) - (3470 - 3300)$$
$$= 240 - 170 = +70 \text{ cm}^{-1}.$$

This is a relatively large positive value, due to interaction between the fast F-H stretching mode and the slow O ... F stretching mode. (Without the existence of the H-bond $X_{1\sigma}$ would be, of course, zero.) This coupling constant is the key quantity of all H-bond theories based on the interaction of these two motions.

Still in 1965 Le Calvé, Grange and Lascombe [39] independently presented the gas phase IR spectra of dimethylether with HF and DF and extended them to 300 cm^{-1}. They found the two components of v_b, the bridge deformation of higher frequency, at 750 and 660 cm^{-1} for the HF complex and at 550 and 490 cm^{-1} for the DF complex. They also studied the bands of $(HF)_n$ polymers and 1:n complexes which are obtained at higher acid pressures.

Couzi, Le Calvé, Huong and Lascombe [40] remeasured the IR spectrum of $(CD_3)_2O.DF$ and found 2555 cm^{-1} for the central peak and 2770 and 2400 cm^{-1} for the side bands. With these values the $X_{1\sigma}$ anharmonic coupling constant is:

$$X_{1\sigma} = (2770 - 2555) - (2555 - 2400) = 215 - 155 = +60 \text{ cm}^{-1}$$

slightly less than for the HF complex.

Couzi, Le Calvé, Huong and Lascombe [40] also investigated the H-bonded complexes formed by HF or DF with methyl iodide, acetonitrile and acetonitrile

Table 1. Characteristic infrared frequencies of complexes of HF and DF with different bases. From M. Couzi, J. Le Calvé, P. V. Huong and J. Lascombe, J. Mol. Struct. 5, 363 (1970). Reproduced by permission from the Elsevier Scientific Publishing Company.

Bases B	Complexes 1:1 FH ... B	
	$\nu(FH)$ (cm^{-1})	$\delta(FH)$ (cm^{-1})
HF monomère	3072	
CH_3I	3800	
CH_3CN	3700–3657–3510	587
CD_3CN	3657	587
CH_3CHO	$\begin{cases} 3640 \\ 3550 \end{cases}$	645
CH_3OH	3530	650
$(CH_3)_2S$	3685–3502–3350	575
$CH_3COOC_2H_5$	$\begin{cases} 3571 \\ 3489 \end{cases}$	695
$(CH_3)_2CO$	$\begin{cases} 3775–3565–3370 \\ 3675–3460–3275 \end{cases}$	695
$(CD_3)_2CO$	$\begin{cases} 3775–3565–3370 \\ 3675–3460–3275 \end{cases}$	700
$(CH_3)_2O$	3700–3460–3300	755–665
$(CD_3)_2O$	3700–3455–3300	755–665
$(C_2H_5)_2O$	3650–3390–3220	770–690
$(iC_3H_7)_2O$	3365	785–690

Bases B	Complexes 1:1 FD ... B	
	$\nu(FD)$ (cm^{-1})	$\delta(FD)$ (cm^{-1})
DF monomère	2907	
CH_3CN	2680	445
	2655	
$(CH_3)_2CO$		510
	2560	
	2655	
$(CD_3)_2CO$		510
	2560	
$(CH_3)_2O$	2557	540–490
$(CD_3)_2O$	2770–2555–2440	555–492

—d_3, methanol, acetaldehyde, dimethylthioether, acetone and acetone —d_6 and ethylacetate from 5000 to 200 cm^{-1}, at a variety of partial pressures. The frequencies of v_1 and v_b in the different systems are listed in Table 1 taken from [40]. Typical for the carbonyl containing acceptors is the case of the acetone. HF complex. The main peak is at 3460 cm^{-1}, very close to its location for ether. HF; three subbands are resolved at higher frequencies, at 3565, 3675 and 3775 cm^{-1} and at 3370 and 3275 cm^{-1}. It is remarkable that the spacing is not of the order of 180 cm^{-1} which would correspond to the known O ... F stretching mode (v_σ); it is closer to 100 or 110 cm^{-1} which very likely corresponds to the bridge deformation v_β. Couzi et al proposed an alternative explanation. According to these Authors the observed bands belong to two different H-bond complexes: one in which the acceptor is the lone pair on the oxygen and one in which it is the π bond of the carbonyl group. The one would be responsible for the intense band at 3460 and the subbands at 3675 and 3275 cm^{-1}; to the other would belong the (apparently) intense band at 3565 and the subbands at 3775 and 3370 cm^{-1}. This is possible and it would interpret the subbands in terms of combinations involving v_σ. The present reviewer, however, favors the first interpretation. We have already encountered cases where the main spacing in the fine structure was determined by the low frequency bridge deformation frequency, v_β. This depends on Franck-Condon and Boltzmann factors and the respective anharmonic coupling constants which may favor v_σ or v_β in given cases. Other observations can easily be made by looking at Table 1.

On the whole these results are in line with those obtained with ether .HF complexes.

3.3 The v(XH) Overtone and Anharmonicity Constants

In order to understand the nature of H-bonding we have to examine the interaction between the fast X-H stretching vibration (v_1) and the slow motions of the bridge. This is essentially a problem of anharmonic interaction and as long as the concept of normal vibrations can be applied to H-bond systems the interaction will result in the appearance of combination bands. Since this is the case of at least weak and medium strong H-bonds gaining knowledge of anharmonicity constants in H-bond systems is of fundamental importance. The most important of these is X_{13} (or $X_{1\sigma}$) which connects the v_1 and v_σ stretching vibrations. It is of interest, however, to know the extent to which the anharmonicity of v_1, itself, X_{11}, is affected by H-bond formation and any numerical data on these and the other anharmonicity and coupling constants constitute much needed input for H-bond theories.

An approximate value for $X_{1\sigma}$ can be deduced from Millen's original spectra. As described in Section 3.2 for the diethylether-hydrogen fluoride system, according to assignments made by Arnold and Millen [20] the main v_1 band is at 3405 cm^{-1}. The two weaker bands at 3655 and 3225 cm^{-1} can be assigned to the summation tone ($v_1 + v_\sigma$) and to the difference tone ($v_1 - v_\sigma$) respectively. It is important to remember in this respect (sect. 2.1) that while the frequency of a difference tone is the simple difference of the frequency of the two fundamentals:

$$(v_1 - v_\sigma) = v_1^{01} - v_\sigma^{01}$$

the summation tone involves the coupling constant

$$(\nu_1 + \nu_\sigma) = \nu_1^{01} + \nu_\sigma^{01} + X_{1\sigma}.$$

In this case $3405 - 3225 = 180 \text{ cm}^{-1}$ is practically the same as the $\nu_\sigma = 175 \text{ cm}^{-1}$ measured directly by Thomas [38]. Then

$$X_{1\sigma} = 3655 - (3405 + 180) = +70 \text{ cm}^{-1}.$$

The positive sign reflects the Sheppard effect. As seen above the same result is obtained from Millen's dimethylether.HF spectrum.

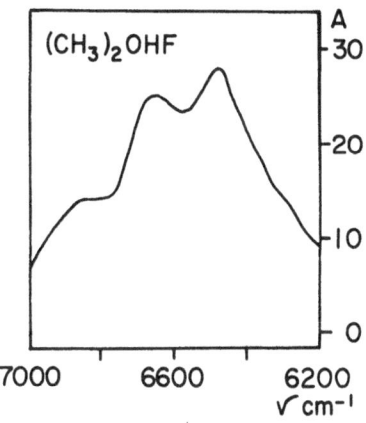

Fig. 7. The overtone of the H-F stretching vibration of the dimethylether-hydrogen fluoride complex in the gas phase. From J. W. Bevan, B. Martineau and C. Sandorfy, Can. J. Chem. 57, 1341 (1979). Reproduced by permisson from the National Research Council of Canada

Table 2. The frequencies of the HF or DF stretching fundamental, first overtone and the anharmonicity constant X_{11}. From J. W. Bevan, B. Martineau and C. Sandorfy, Can. J. Chem. 57, 1341 (1979). Reproduced by permission from the National Research Council of Canada.

Complex	ν_1^{01}	ν_1^{02}	X_{11}
$(CH_3)_2O$-HF	3470	6485	-228 (± 28)
$(CH_3)_2O$-DF	2540	4860	-110 (± 28)
$(C_2H_5)_2O$-HF	3405	6335	-237 (± 28)
$(C_2H_5)_2O$-HF	2500	4780	-110 (± 28)
$(C_2D_5)_2O$-HF	3400	6325	-237 (± 28)
$(C_2D_5)_2O$-DF	2500	4770	-115 (± 28)
$(CH_3)_2CO$-HF	3470	6530	-205 (± 28)

In order to obtain X_{11} we need the first overtone of v_1. This was measured for ether .HF and acetone .HF systems by Bevan, Martineau and Sandorfy [41]. In addition to obtaining X_{11} it was hoped to find subbands of the $(2v_1 \pm v_\sigma)$ type since these must yield the same $X_{1\sigma}$ coupling constants as the subbands of the v_1 fundamental. Figure 7 shows the overtone spectrum of $(CH_3)_2O.HF$ and the relevant data for this and other similar systems are collected in Table 2.

The first observation is that these spectra resemble Millen's original spectra in the v_1 fundamental region so that the results obtained in the fundamental and in the the overtone regions mutually corroborate each other. The rest depends on the assignments of the subbands at both levels. A detailed discussion of these was given in [41] and will not be repeated here. The most likely interpretation of the spectra is in line with Millen's original assignments for this system for v_1^{01}. Taking for $(CH_3)_2O.HF$ the 6485 cm^{-1} band for the "pure overtone" $((0, 0) \rightarrow (2, 0)$ or $v^{02})$ we obtain $X_{11} = -228$ cm^{-1}. It is somewhat above -200 cm^{-1} for all the complexes studied. Since the free HF vibrator has an X_{11} value of -90 cm^{-1} this corresponds to a greater than twofold increase due to H-bond formation. The ether .DF complexes have X_{11} values of about -110 cm^{-1} which can be compared to the "free" value of -45 cm^{-1}. (To the second order the H/D ratio of the anharmonicity constants is equal to the ratio of the reduced masses which is about 2.) As was shown in [41] there exists another, less probable, assignment of the v_1^{02} which is compatible with the observed data. This would put the "pure overtone" to the subband of 6650 cm^{-1} for $(CH_3)_2O.HF$, 6525 cm^{-1} for $(C_2H_5)_2O.HF$ and 6660 cm^{-1} for acetone .HF. Then we obtain about -140 or -145 cm^{-1} for X_{11}. Anyway there is no doubt about the substantial increase of X_{11} upon H-bond formation. Just as far the fundamentals deuteration of the *alkyl* groups leaves the overtone region practically unchanged.

For the OD complexes the overall shape of the overtone group of bands exhibits significant differences with respect to the OH complexes. This is a situation similar to that encountered in the fundamental region. The observed subbands cannot, in general, be assigned to $(2v_1 + v_\sigma)$ type combinations; one must involve $(2v_1 + v_\beta)$ type combinations involving bridge deformation vibrations. It is, indeed, possible that the peaks which seem to be the "pure" fundamental or overtone are actually a combination of the $(v_1 + v_\beta)$ or $(v_1 + 2v_\beta)$ and $(2v_1 + v_\beta)$ or $(2v_1 + 2v_\beta)$ types respectively. For this reason it is perhaps better not to use the OD frequencies for computing anharmonicities until better resolved spectra become available.

Despite of the difficulties encountered some conclusions may be drawn. The overtone spectra are similar to those which Millen originally obtained in the fundamental region. H-bond formation causes a substantial increase in the value of the anharmonicity of the X-H stretching vibration. The coupling constant between the fast and slow stretching motions also has a substantial value and supplies a much needed parameter for H-bond theories.

Several other gas phase overtone data became available through the investigations of Bernstein and his coworkers [42]. Some of them are shown in Table 3. It is seen that except for the weakest H-bonds the X_{11} values undergo a substantial increase upon H-bond formation just as for the ether .HF systems. In all cases just as for ether .HF the bands are broad; the numbers are to be considered as good approximate values.

Table 3. Frequencies of the OH or OD fundamental, first overtone and the anharmonicity constant X_{11}. Unpublished work by H. J. Bernstein, D. Clague, A. Gilbert, A. J. Michel and A. Westwood. Reproduced by permission from the Authors.

	Free			Associated		
	v_1^{01}	v_1^{02}	$-X_{11}$	v_1^{01}	v_1^{02}	$-X_{11}$
$CF_3CH_2OH.O\ Et_2$	3660	7150	85	3443	6740	73
$(CF_3)_2CHOH.O\ Et_2$	3650	7130	85	3336	6370	151
$(CF_3)_3COH.O\ Et_2$	3632	7098	83	3152	5860	222
$CF_3OH.NH_3$	3684	7201	84	3511	6835	94
$CH_3CH_2OH.NH_3$	3660	7148	86	3350	6318	191
$CH_3OH.N(CH_3)_3$	3684	7201	84	3365	6292	210
$C_2H_5OH.N(CH_3)_3$	3672	7169	88	3348	6288	203
$(CH_3)_2CHOH.N(CH_3)_3$	3652	7146	79	3333	6318	174

3.4 Systems With Polyatomic Proton Donors

In the preceding sections complexes of ethers with hydrogen chloride and hydrogen fluoride were dealt with to some length. The logical next step is to ascertain if the essential results obtained on these relatively simple systems can be generalized to more complicated ones where both the proton acceptor and the proton donor are polyatomic. Still in 1965 Millen and Zabicky investigated amine-alcohol systems [24] and Millen and Samsonov [23] ether-nitric acid systems. In a second series of papers, from 1974 to 1979 Millen and his coworkers examined amine-alcohol [43,44], phenol-trifluoro-ethanol [45], alcohol-ether, chloroform-amine [46], O-H ... N complexes [47] and hydrogen fluoride-alcohol systems [48]. Tucker and Christian [49] investigated the alcohol-trifluorethanol complex. In general, to find the right partners, it is appropriate to pair weak proton donors with stronger proton acceptors like alcohols + amines or stronger proton donors with weak proton acceptors like ethers + HF. If both constituents are "weak" the complex will dissociate in the gas phase; if both are strong a solid complex will be formed with very low vapor pressure.

The qualitative features of the spectra are in conformity with the observations made on the ether .HF or ether .HCl systems. Millen and Zabicky [24] examined the complexes of methanol with trimethylamine, dimethylamine, methylamine, ammonia and aziridine. In all cases methanol is the proton donor and the nitrogen lone pair is the electron acceptor. In the methanol trimethylamine-complex spectrum the central band is at 3355 cm^{-1}. One subband is resolved at each side, at about 3495 and 3200 cm^{-1}. Using a large excess of amine the absorbance of the bands is proportional to the product of the pressures of the two components so the spectrum must be attributed to the 1:1 complex. From the spacing a value of 145 cm^{-1} can be infered for v_σ.

The complexes of methanol with the other amines gave similar spectra. The distance (Δv) from the free band varies with base strength:

	Me$_3$N	Me$_2$NH	MeNH$_2$	NH$_3$	$\overset{NH}{\underset{CH_2-CH_2}{\diagup\diagdown}}$
v_1	3350	3380	3445	3510	3460
Δv	330	300	235	170	220

It can be taken as an approximate measure of electron donor ability of the bases. Several years later Hussein and Millen [43] carried out a more extensive study with a variety of alcohol-amine complexes. This time they found that the electron donating abilities of the amines decreased in the following series:

$$Et_3N > Et_2NH \approx Me_3N > Me_2NH > EtNH_2 > MeNH_2 > NH_3 .$$

The spectra could be always interpreted in terms of $v_1 \pm v_\sigma$ sum and difference bands. For any given alcohol Δv decreases along the series Me$_3$N > Me$_2$NH > > MeNH$_2$ > NH$_3$.

Subsequently Millen and Mines [44] determined the thermodynamic parameters for amine-methanol systems from pressure, volume and temperature measurements in the gas phase. The equilibrium constants for the formation of the complexes were determined over the temperature range 25–45 °C. The ΔH values correlate well with the H-bond shift Δv of the O-H stretching bands. Both can be considered as approximate measures of the electron donating ability of the amines. Table 4 is taken from reference [44]. The ΔH^0 for the trimethylamine-methanol complex were previously measured by Fild, Swiniarski and Holmes who obtained similar values [50]. It is interesting to note that the ΔH are somewhat *higher* in the gas phase than in solution meaning that the sum of the solvation energies of the two monomers, methanol and amine, is larger than the solvation energy of the complex.

Table 4. Comparison of ΔH and Δv for gas-phase amine-methanol complex formation. From D. J. Millen and G. W. Mines. J.C.S. Faraday Trans. II. *70*, 693 (1974). Reproduced by permission from the Chemical Society.

COMPLEX FORMATION

amine	Et$_3$N	Et$_2$NH	Me$_3$N	Me$_2$NH	MeNH$_2$
$-\Delta H$/kJ mol^{-1}	31.4	28.0	28.9	25.9	23.4
Δv/cm^{-1}	370	325	310	295	245

Hussein, Millen and Mines [45] studied the spectra of a number of H-bonded complexes formed between donors more acidic than the previous ones: 2,2,2-trifluomethanol and phenol, with a variety of oxygen containing acceptors. The spectra are similar but yield less information than those of the complexes discussed so-far. As can be inferred from the Δv values the electron donating power of the oxy group increases along the series esters, ketones and aldehydes, alcohols and ethers, and for any given class it increases by increasing methylation.

Complexes formed by trifluomethanol or phenol with ketones and esters generally show two maxima. For example, for the acetone-trifluoromethanol complex they are at 3527 and 3454 cm^{-1}. They were interpreted [45] as being due to two different complexes: O-H ... π and O-H ... O as has been assumed for carbonyl compounds by Couzi et al. [40]. Insensitiveness to pressure ratios then shows that both must be 1:1 complexes. Whether or not this interpretation could be maintained if better resolved spectra became available is yet to be demonstrated, however.

Tucker and Christian [49] determined the thermodynamic parameters for the acetone-trifluomethanol complex. It turned out to be significantly more stable in the gas phase than in CCl$_4$ solution.

The weakest complexes studied in the gas phase were alcohol-ether and chloroform-amine systems. (Hussein and Millen [46]). For the order of increasing basicity they obtained, for the ethers:

$$Me_2O \approx 1,4\text{-dioxan} < Et_2O < \text{tetrahydrofuran} < \text{diisopropylether}.$$

For the amines they found $NH_3 < MeNH_2 < Me_2NH < Me_3N$. For the chloroform-amine complexes the Δv are 12, 26, 45 and 53 cm^{-1} in this order. No subbands were resolved. This is, no doubt, due to the low frequency of the bridge stretching vibration v_σ for these very weakly H-bonded systems and to the weakness of the anharmonic coupling between them. The existence of the molecules in more than one conformation broadens the bands and makes the observation of the subbands even more difficult.

Later Legon, Millen and Schrems [48] determined the gas phase IR spectra of complexes of a series of alcohols with hydrogen fluoride. With appropriate pressure ratios 1:1 complexes are obtained, HF being the proton donor. Supposing again that the Δv values are a good measure of the electron donating power of the oxygen lone pairs in the free alcohols the following series is obtained:

$$CH_3OH < CH_3CH_2OH < (CH_3)_2CHOH < (CH_3)_3COH .$$

The Δv value increases by steps of about 25 cm^{-1} in this order. Electron attracting substituents decrease the electron donating power of the oxygen atom. The v_1 frequencies are, for the H$_2$O, CH$_3$OH and (CH$_3$)$_2$O complexes of HF, 3608, 3530 and 3470 cm^{-1} respectively.

For the methanol-HF complex the associated HF band v_1 has its center at 3530 cm^{-1}. Unfortunately the $(v_1 + v_\sigma)$ summation band is overlapped by the methanol OH band. The difference band is at 3340 cm^{-1} so that v_σ can be estimated to be 190 cm^{-1}. With DF v_1 moves to 2602 cm^{-1}. All this is in line with the observation reported on other HF complexes in Section 3.2 and does not supply essentially new information on the nature of H-bonding.

An interesting system is the ether-nitric acid system, studied by Millen and Samsonov [23], Carlson, Witkowski and Fateley [51] and Al-Adhami and Millen [52]. The main v_1 band is at 2950 cm^{-1} with subbands at 3125 and 2760 cm^{-1} and a shoulder near 2600 cm^{-1}. From the spacing and from the far IR a value of 185–195 cm^{-1} is obtained for v_σ for dimethylether.

Extensive gas phase work on alcohols was later reported by Barnes, Hallam and

Jones [53-54] who used path lengths ranging from 9 m to 40 m and by Reece and Werner [55]. Interestingly, the former observed a number of combination bands at both sides of the monomer v_1 band of 2,2,2-trifluoroethanol (TFE) and some other fluoro-alcohols. After considering several alternative explanations Barnes et al. [53] arrived at the conclusion that these combination bands belong to the monomer and that their unusual strength is probably due the existence of O-H ... F, intramolecular H-bonds within these fluoro-alcohols (cf. Krueger and Mettee [56,57]). They presented complete vibrational assignments for TFE. The value of the OH torsional vibration, 280 cm^{-1}, should be mentioned in this context. The alkanols did not exhibit such combination bands for the monomer except with the OH torsional mode. Inskeep, Kelliher, McMahon and Somers [58] already described dimer alcohol bands in the gas phase. Barnes et al. [53] determined v_1 values, the corresponding Δv from the free band position and ΔH values for many such systems. Others have been reported by Inskeep et al. [58], Clague, Govil and Bernstein [59], Kivinen and Murto [60] and Murto et al. [61]. The observed OH torsional frequencies are 286 (a_1) and 205 (e) for methanol, 201 for ethanol.

In their second paper Barnes et al. [54] studied heteroassociations of alcohols in the gas phase at low alcohol pressures (about 2 Torr) and much higher pressures of the proton acceptors. In particular the IR spectra were measured for TFE with acetonitrile, acetone, methyl-isocyanide, diethylether and tetrahydrofuran and a number of amines. The dimers were obtained in all cases. Their paper contains a number of v and Δv values for such systems.

Reece and Werner [55] measured the Δv shifts from the free OH position for a number of complexes of alcohols with a number of proton acceptors both in the gas phase and in CCl$_4$. They found that for small shifts (weak complexes) the ratio of the frequency shift in the vapor to that in CCl$_4$ is about 0.816. Table 5 shows some of their results.

Table 5. Monomer frequencies (cm^{-1}) and frequency shifts (Δv cm^{-1}) for association in the vapour phase at 27 °C.
T.F.P. = 2,2,3,3-tetra-fluoro propan-1-ol. From I. H. Reece and R. L. Werner, Spectrochim. Acta 24A, 1271 (1968). Reproduced by permission from Pergamon Press, London.

Alcohol	t-butanol	ethanol	methanol	propargyl	T.F.P.
Monomer	3646	3675	3688	3663	3658
Acceptors					
Acetaldehyde					90
Acetone	96		90	108	125
Methyl acetate	65		66		108
Diethyl ether	116	129	133	178	210
Tetrahydrofuran	115	118	124	174	210
p-Dioxane			100		184
t-Butanol			115		180
Methanol			88		156
Acetonitrile			58		103
Pyridine			210		370

3.5 The Cyclic Dimers of Carboxylic Acids

The IR spectra of carboxylic acids can be measured in the vapor phase with relative ease. They exist there as either monomers or cyclic dimers depending on pressure and temperature. The first vapor phase spectra of formic acid and acetic acid were published as long ago as 1953 and 1956 by Hadži and Sheppard [62] and by Bratos, Hadži and Sheppard [80]. The main ν_1(OH) stretching region is centered at about 3000 cm^{-1} and has a half-width of about 500 cm^{-1}. It is broad and contains many subbands. The carbonyl frequency is around 1720 ± 20 cm^{-1} with a half-width of about 50 cm^{-1}. The corresponding monomer bands are at about 3500 and 1770 ± 30 cm^{-1} respectively. The vapor, solution and solid spectra do not differ much from one other. Both oxygen and carbon deuterated acids were used to help making vibrational assignments. The main ν_1 band for the OD compounds has its center near 2300 cm^{-1} with a half-width of about 200–250 cm^{-1}.

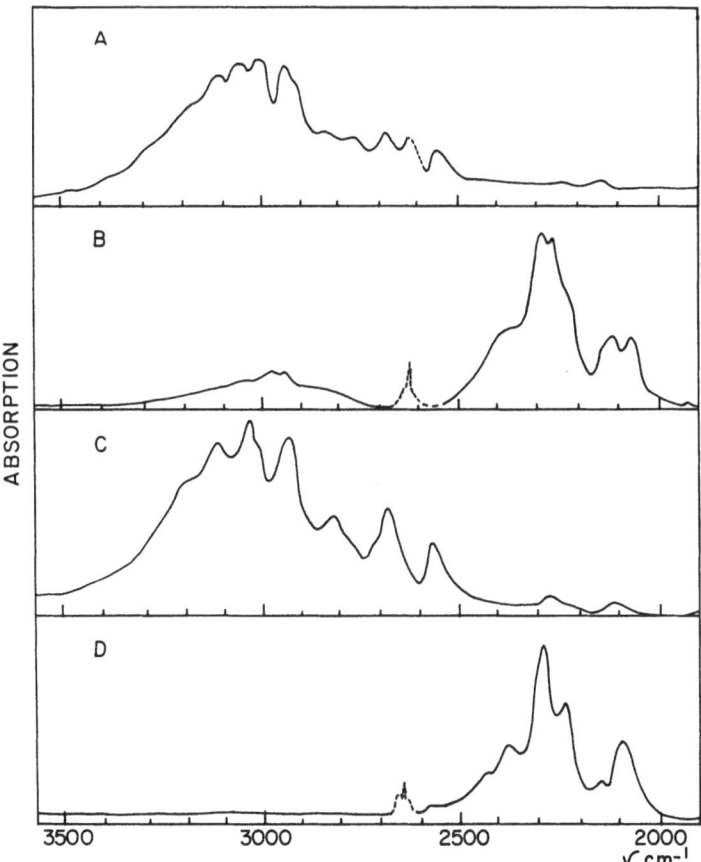

Fig. 8. Infrared spectra of (A) CH$_3$COOH; (B) CH$_3$COOD; (C) CD$_3$COOH; (D) CD$_3$COOD in the gas phase at room temperature. From M. Haurie and A. Novak, J. Chim. Phys. 62, 146 (1965). Reproduced by permission from the Société de Chimie-Physique

Table 6. Bands in the vibrational spectra of the cyclic dimer of CH_3COOH. VS = very strong; S = strong; m = medium; w = weak; vw = very weak; sh = shoulder. The numbers in brackets represent intensities on a relative scale. P = polarized. (c) = bands not belonging to the cyclic dimer. (d) = coupled vibrations. From M. Haurie and A. Novak, J. Chim. Phys. 62, 146 (1965). Reproduced by permission from the Société de Chimie-Physique.

Infrared			Assignment	Raman	Assignment
Gas	Solution	Liquid		Liquid	
3400 sh					
3185 sh	3179 sh	3200 sh	$vC = O$, IR + $vC = O$, R		
3130 S	3098 sh	3092 sh	$vC = O$ IR + δOH R		
3072 S			$vC = O$ IR + $\delta_a CH_3$R		
3027 S	3028 S	3028 S	vOH	3032 (2)	$v_s'CH_3$
3010 sh	2998 sh		$vC = O$ IR + $vC - O$ R	3000 (2)	$v_a CH_3$
	2975 sh				
2958 S	2922 S	2927 S	$vC - O$ IR + $vC = O$ R	2949 (10) P	$v_s CH_3$
2925 sh					
2845 m	2837 sh	2847 sh	δOH IR + δOH R $\delta_s CH_3$ IR + $\delta_s CH_3$R	2865 (0)	$2\delta_s'CH_3$
2777 m	2776 sh	2776 sh	δOH IR + $\delta_s CH_3$R δOH IR + $vC - O$ R		
2698 m	2680 m	2675 m	$vC - O$ IR + $\delta_s CH_3$R		
2640 m	2625 m	2625 m	$vC - O$ IR + $vC - O$ R		
2562 m	2550 m	2550 m	$\delta_s CH_3$IR + $vC - C$ R		
2250 vw		2242 vw	$vC - O$ IR + $vC - C$ R		
2160 vw			$vC - C$ IR + $vC - C$ R (c)		
	1779 sh	1780 sh		1765 sh	
	1758 vw	1758 m		1713 sh	
1730 vs	1712 vs	1715 vs	$vC = O$ (c)	1675 (1) P	$vC = O$
		1650 sh			
1510 vw	1509 vw	1514 vw	δCOO IR + $vC - C$ R		
1422 m	1415 m	1413 m	$\delta_a CH_3$ $\delta_s CH_3$ δOH (d)	1436 (1) P	$\delta_a CH_3$ $\delta_s CH_3$ δOH (d)

1365 sh		1359 m	δ,CH₃	1370 (1) P	δ,CH₃
1292 S		1295 S	vC – O (d)	1283 (0)	vC – O (d)
		1240 sh	(c)		
1060 vw		1050 vw	ϱ,CH₃	1018 (0)	ϱ,CH₃
1005 m		1013 vw	ϱ,CH₃		
940 m		934 m	γOH		
890 sh		886 vw	vC – C	886 (5) P	vC – C
				869 sh	(c)
			δCOO		δCOO
621 m		624 m		624 (2)	
			γCCO		γCCO
		600 sh	(c)	600 sh	(c)
475 m		480 m	δCCO	448 (1)	δCCO
		455 sh	vₛ		

69

The origin of the satellite bands was throughly discussed by Bratos, Hadži and Sheppard [8]. They rightly recognized that "... (they) represent summation freqoencies involving lower frequency fundamentals of the COOH(COOD) groups, perhaps brought up in intensity by Fermi resonance with the νOH (νOD) vibration." Later Haurie and Novak [63,64] carried out a new study on CH_3COOH, CH_3COOD, CD_3COOH and CD_3COOD in the gas phase, in the liquid phase and in solution from 4000 to 400 cm^{-1}. Their gas phase spectra are shown in Fig. 8. A part of their table (Table 6) listing the observed frequencies of the cyclic dimer and their assignments is also reproduced. A few comments can be made at this point.

Since the cyclic dimer possesses a center of symmetry all bands allowed in the IR are forbidden in Raman and vice-versa.

Unfortunately the Raman active νOH(OD) frequency could not be identified in the gas phase, for reasons well known to spectroscopists. (The OH bond is too polar and not sufficiently polarisable.) Haurie and Novak measured the Raman spectrum in the pure liquid state and saw a weak band at 3032 cm^{-1} which they assigned to νOH while the IR value is 3027 in the gas phase and 3028 in the liquid. For CD_3COOH the Raman frequency they found is 2985 cm^{-1} and the IR frequency is 3040 in the gas phase and 3028 in the liquid. The IR and Raman values differ astonishingly little.

Since a near coincidence of the a_g and b_u ν(OH) bands is not expected we have to conclude that the a_g band has not yet been seen. Indeed, the IR and Raman carbonyl frequencies are split by 40 cm^{-1}, 1715 (b_u) — 1675 (a_g) for liquid CH_3COOH, by 64 for CH_3COOD, 53 for CD_3COOH and 57 cm^{-1} for CD_3COOD despite the fact that carbonyl frequencies are always much less altered by H-bond formation than OH frequencies. Even the in-plane OH bending band (δOH in [63]) is split by 23 cm^{-1} for CH_3COOH and by 22 cm^{-1} for CD_3COOH. Several years later Foglizzo and Novak [65] measured the Raman spectra of acetic acid in the polycrystallin solid state and identified the ν(OH) band as a weak band at 2900 cm^{-1} for CH_3COOH and 2910 for CD_3COOH. In this physical state, however, the acidic acid molecules form infinite H-bonded chains, not cyclic dimers. The lack of knowledge of the second ν(OH) band is a serious drawback and hampers, among others, overtone studies on carboxylic acid dimers.

Next we have to ask: where are the low frequency bridge vibrations of carboxylic acids (like ν_σ and ν_β, well known for alcohols) and do they appear in the fine structure of the ν(OH) stretching band?

Costain and Srivastava [66,67] measured the microwave rotation spectrum of CF_3COOH with three different partners: HCOOH, CH_3COOH and CH_2FCOOH. The results of this important work will be reported in Section 5. Here we only mention that they inferred a vibrational satellite from the pure rotational spectrum with a frequency of 180 cm^{-1} which can be assigned to one of the bridge stretching frequencies, ν_σ. All other available data seem to relate to the crystallin phase or to solutions. In solution, however, most carboxylic acids from cyclic dimers so that such data can be compared to vapor data. The following are from Jacobsen, Mikawa and Brasch (Table 7) [68,69]. Data for other carboxylic acids are found in references [68] and [69]. For the higher acids the frequencies tend to be lower than for formic- and for acetic acid.

These frequencies are in the same range as, for example, for the ether. HF complex and are not extraordinarily high. Then one would expect these frequencies to show up among the subbands of the vOH stretching band, both as summation and as difference bands. Furthermore hot bands of the $v_1 + (n'v - n''v)$ type should contribute to the intensity and breadth of the complicated band system around v_1. As Haurie and Novak [63,64] has shown, however, all the observed peaks can be accounted for as combinations of internal modes not involving the bridge vibrations. Also the temperature effect characteristic of hot bands has not so far been observed. In the writer's opinion the problem is still open. In view of the breadth of the v_1 system many hidden bands could be present and many coincidences are possible. All this, however, does not have to shaken our ideas about H-bonding. The appearence of combination bands, either those involving bridge vibrations or those involving internal modes is a matter of Franck-Condon factors in the Sheppard-Stepanov sense, the closeness of coincidences for Fermi resonance, the values of anharmonic coupling constants. We have no detailed knowledge about these. It is quite possible that in ether .HCl or ether .HF, etc. complexes combinations involving the bridge modes are predominant whereas in carboxylic acid dimers combinations of internal modes are more prominent as it appears to be the case. (Witkowski and Maréchal [37] take a different view of this problem. See also the review paper by Maréchal [101].)

Table 7. The low frequency bridge vibrations for the cyclic dimers of formic acid and acetic acid. Data from Jacobsen, Mikawa and Brasch [68-69].

Formic acid	Vapor	Solution
$v_\sigma(OH ... O)$	248	248
bridge deformation modes	164	173
twist	68	

For $(HCOOD)_2$ and $(DCOOH)_2$ closely similar frequencies were found.

Acetic acid	Vapor	Solution
$v_\sigma(OH ... O)$	188, 168	176
twist	50	

Vapor phase spectra of trifluoro-acetic acid were reported by Christian and Stevens [70] and of oxalic acid by Pava and Stafford [71].

4 Information Derived From the Rotional Fine Structure of Vibrational Bands

In the preceding sections vibrational spectra were used as a source of information on H-bonding. The rotational fine structure of the bands could not be resolved. Only the rotational contours, the breadth of unresolved P, Q, R branches entered occasionally the discussion. In order to resolve rotational fine structure much lower pressures and therefore longer path-lengths are needed. This puts an additional limitation on the selection of systems that can be studied: our complexes must remain complexes at the lower pressures that are required.

The first attempt to resolve rotational fine structure on the IR bands of H-bonded systems has been due to Jones, Seel and Sheppard [74]. They studied the complexes $H_3N \ldots HCN$, $D_3N \ldots DCN$ and the dimer of hydrogen cyanide $(HCN)_2$. Three *parallel* bands were observed for these complexes: the C-H (or C-D) stretching band, the $C \equiv N$ stretching band and the NH_3 (or ND_3) symmetric deformation band. For the ammonia-hydrogen cyanide complex these bands are at at 3150, 2085 and 1040 cm^{-1} respectively.

The $vC \equiv N$ band had a P-R contour with a separation of 13.2 cm^{-1} at 0 °C. The rotational constant, B_v, computed from it through the Gerhard-Dennison formula [75] is 0.113 cm^{-1} (Cf. Sect. 2.4). Assuming that the N ... HCN system is linear and that NH_3 and HCN have the same dimensions in the complex as in the isolated molecules, the N ... C distance is calculated to be 2.96 ± 0.1 A, likely to imply a slight lenghtening of the C-H bond. The vC-H band at 3150 cm^{-1} has a half-width of about 80 cm^{-1}. Several sharp bands appear at its low frequency side having a half-width of about 5 cm^{-1}. (At 3158, 3150, 3138, 3126, 3117 and 3107 cm^{-1}.) The average spacing of these sharp bands is 10.2 cm^{-1}. No $(v_1 \pm v_\sigma)$ bands, similar to those in Millen's ether .HF or ether .HCl complexes has been found.

The $D_3 \ldots DCN$ complex has the related bands at 2560, 1890 and 805 cm^{-1}. The vC-D band has a half-width of about 60 cm^{-1}, no fine structure could be resolved.

Under the given experimental conditions the $(HCN)_2$ dimer gave only one band, at 2095 cm^{-1} $(vC \equiv N)$. This band had a P-R separation of 9.2 cm^{-1}. The rotational constant, B_v, was found to be 0.057 cm^{-1}, leading to a C ... N bond length of 3.34 A. The vC-H bands of the dimer were obscured by monomer bands.

The perpendicular bands could not be detected in either complex. (For possible reasons of this see [74]).

As to their most important observation, the appearance of sharp bands on the low frequency wing of the vC-H band of the ammonia · hydrogen cyamide complex Jones, Seel and Sheppard [74] proposed that these can be interpreted as "... a hot band sequence of a type ... where low frequency fundamentals can interact strongly with those of a higher frequency." Although no detailed assignments could be given at the time, this turned out to be a sound idea.

Important advances were made by Thomas and Thompson [76] and by Thomas [77-79] between 1970 and 1975. The first system they examined was the complex formed between acetonitrile (or CD_3CN) and HCl (or DCl). A broad band is observed in the range 2750 to 2700 cm^{-1} indicating a weak complex. (The free HCl band is at 2885 cm^{-1}.) The band has a half-width of about 100 cm^{-1}. On the broad contour a number of sharp peaks are superimposed at the lower frequency side (Fig. 9). A weaker band corresponding to $(v_1 - v_\sigma)$ is also observed. The $(v_1 + v_\sigma)$ summation band probably contributes to making the shape of the main band asymmetrical but could not be clearly identified. Several other complexes between alkylcyanides and HCl (or DCl) were studied. The v_σ bridge frequency is 100 cm^{-1} for CH_3CN, 95 cm^{-1} for C_2H_5CN, 90 for $CH_2 = CHCN$ and 85 for tert-C_4H_9CN as estimated from the separation between v_1 and $(v_1 - v_\sigma)$. The most important feature is the series of sharp bands that is observed on the low frequency wing of v_1. About nine bands are listed by Thomas and Thompson. The average spacing among these fine bands is a little as 1.33 cm^{-1} for $CH_3CN.HCl$ and 0.94 cm^{-1} for $CH_3CN.DCl$.

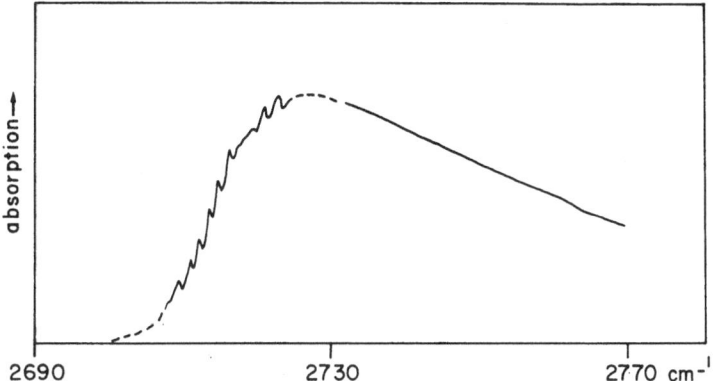

Fig. 9. The vH-Cl band of the $CH_3C \equiv N.HCl$ complex in the gas phase. From R. K. Thomas and Sir. .H. Thompson, Proc. Roy. Soc. London A316, 303 (1970). Reproduced by permission from the Royal Society

The complex is a symmetric top with C_{3v} symmetry. The moment of inertia about the C_3 axis, I_A, is very much smaller than the moment of inertia about the axis perpendicular to the axis, I_B, so it is to be expected that the rotational fine structure will consist of P and R branches with a much weaker and hardly noticeable Q branch. Now, the observed sharp lines cannot be individual P lines, these would be much more closely spaced. So Thomas and Thompson [76] suggested that they are due "to a conglomeration of lines of each of a series of hot bands." This can be rationalized as follows. We suppose that the H-bond is shorter in the $v = 1$ state of v_1 than in the $v = 0$ state (The Sheppard effect.) Then B′, the rotational constant for $v = 1$ will be larger than B″, the rotational constant for $v = 0$. Then B″ − B′ will be negative and if it has a sufficiently large value there will be band heads in the P branches. (Cf. Herzberg [18], pp. 106–115). (The much smaller rotational constant about the axis perpendicular to C_3 may be disregarded.) Furthermore, the rotational lines approaching the band head will be "so close as to coalesce into a pronounced maximum in the contour." This applies to the whole sequence of close-lying hot bands which contribute to the apparent intensity of v_1. (Brown and Sheppard [80], Jones. Seel and Sheppard [74].) Then, as Thomas and Thompson stated, the sharp peaks have to be "interpreted as being due to an agglomeration of P lines near the band head in each of a series of hot bands which accompany that association band, and which arise from transitions from higher levels of a bending mode of low frequency." The latter point needs some elaboration. One might think that the hot bands concerned arise from a series of excited levels of the H-bond stretching vibration, v_σ: $(v_1 + n'v_\sigma - n''v_\sigma)$. The frequency of this vibration, about 100 cm^{-1}, would, however, make the population of the excited levels too low and fall off more rapidly in higher levels than what is observed. On the other hand the bridge bending frequency v_β, with an estimated frequency between 20 and 60 cm^{-1} would fit the observed spectrum. Other low frequency modes might contribute to shaping the band contour.

This argument underscores the important role played by the low frequency H-bond bending mode for determining the shape and width of the main v_1 band. One will recall in this respect the additional argument put forward by Bertie and Falk [28] which was mentioned in Section 3a. Still another piece of supporting evidence for this was found by Bevan, Martineau and Sandorfy [41] who did gas phase work on the first overtone of v_1 of the ether .HF complex (Sect. 3.3). Theoreticians should take account of this in their calculations.

Subsequently Thomas [77] examined four other typical complexes in a similar manner: ether .HF, HCN.HF and CH$_3$CN.HF, and H$_2$O.HF. For ether .HF he first determined the enthalpy of formation of the H-bond using v_1 and the Benesi-Hildebrand method [81]. Here are some of the results he obtained.

Complex	Δv_1/cm^{-1}	ΔH/kJ mol^{-1}
$(CH_3)_2O.HF$	505 ± 10	-43
$CH_3OC_2H_5.HF$	535	-37
$(C_2H_5)_2O.HF$	575	-30

As is seen the ΔH do not follow the Δv. In this respect Thomas pointed out that the ethylethers may exist in more than one conformation and that the conformation may change when a H-bond is formed. The existence of these different conformations may also be one cause of the broadening of the observed bands.

In the far IR a band was found at about 180 cm^{-1} in all cases which can be assigned to v_σ:

Ether	v_σ with HF	v_σ with DF
dimethyl	185	185
methyl ethyl	180	170
diethyl	175	165

For the dimethylether .HF complex Thomas also recorded the spectrum from 50 to 8 cm^{-1} and found rotational contour around 10 cm^{-1} belonging to the complex meaning that the complex has a lifetime of at least three picoseconds.

The next work by Thomas [78] has been a successful application of the same ideas to the HCN.HF and CH$_3$CN.HF complexes. For these systems all the bands due to bridge vibrations could be identified. For HCN the bridge bending motions are at 555 (v_b) and 70 cm^{-1} (v_β). The v_1 band is centred near 3710 for HCN.HF and 3750 cm^{-1} for CH$_3$CN.HF. Extensive rotational fine structure was found for both systems. For HCN.HF the average spacing is 4.3 cm^{-1}, for CH$_3$CN.DF it is 2.7 cm^{-1}; for DCN.DF it is 2.7 and for CH$_3$CN.DF 1.8 cm^{-1}. Fine structure was also found on the $(v_1 - v_\sigma)$ difference band with an average spacing of 2.9 cm^{-1} for CH$_3$CN.HF. Less well resolved fine structure was found on v_b with spacing of 4 cm^{-1} for CH$_3$CN.HF and 7 cm^{-1} HCN.HF. The rotational lines were broader in this case.

The frequencies of the fundamental vibrations could be quite accurately determined for these HF-cyanide complexes. They are tabulated in Table 8 taken from reference [78].

Table 8. Fundamental vibrations in (cm^{-1}) of cyanide HF complexes. From R. K. Thomas, Proc. Roy. Soc. London *A325*, 133 (1971). Reproduced by permission from the Royal Society.

complex	A_1 or Σ species		E or Π species	
	$\nu_1(\pm 2)$	ν_σ	$\nu_b(\pm 3)$	$\nu_\beta(\pm 35\%)$
HCN-HF	3710	155^c	555	70
DCN-DF	2720	—	416	62
CH$_3$CN-HF	3627	168^b	620	40
CD$_3$CN-HF	3627	165^b	620	37
CH$_3$CN-DF	2667	160^a	463	39
CD$_3$CN-DF	2668	160^a	461	36

Errors: a $\pm$ 20 cm^{-1}; b $\pm$ 3 cm^{-1}; c $\pm$ 10 cm^{-1}.

As before the subbands right and left from ν_1 were interpreted as ($\nu_1 \pm \nu_\sigma$) combination bands and the fine structure as a series of hot bands of the ($\nu_1 + n'\nu_\beta - n''\nu_\beta$) type. The hot bands are parallel bands with strong P and R branches and very weak Q branches. Since the H-bond shortens in the $v = 1$ state of ν_1, $B' > B''$ and the band heads are in the P branch. Many close-lying rotational lines accumulate around the band head so that what we observe is essentially a series of P branches each one belonging to a hot band. The latter form a sequence in the vibrational quantum number of the bridge bending vibration ν_β.

The fine structure of ν_b (centred at 555 cm^{-1} for HCN.HF and 620 cm^{-1} for CH$_3$CN.HF) is different since this is a perpendicular (degenerate bending) vibration. It has prominent Q branches which again belong to hot bands in ν_β.

According to Bertie and Millen [19] the appearance of the ($\nu_1 \pm \nu_\sigma$) bands is due to the cubic potential constant k_{113} or, according to the notations used here, $k_{11\sigma}$. The coupling constants for the hot bands are equal to their spacing (see Sect. 2.1) that is $4.7 - 1.7$ cm^{-1} for the different complexes for ν_1 and $7.6 - 3.0$ for ν_b. These may seem to be small but they suffice to cause the fine structure and the breadth of the observed vibrational bands!

The 1:1 complex of water and hydrogen fluoride was also studied by Thomas [79], from 4000 to 400 cm^{-1}. This is an experimentally difficult task in view of the low volatility of the H$_2$O.HF complex. The analysis of the spectrum shows that the complex is coplanar, C_{2v}. This splits the degeneracy of ν_b and ν_β the two bridge deformation vibrations which are degenerate in the linear or C_{3v} complexes. ν_1 has some structure consisting of a sharp band at 3608 cm^{-1}, a broader band split into two at 3623 and 3626 and a broad band at 3644 cm^{-1} followed by continuous absorption (Fig. 10). The free-associated separation is 354 cm^{-1} for H$_2$O.HF while it is 420 cm^{-1} for dimethylether .HF. (Arnold and Millen [20].) As in the previous cases the fine structure can be interpreted as a series of hot bands, ($\nu_1 + n'\nu_\beta - n''\nu_\beta$).

The low frequency bridge bending vibration v_β is not degenerate in this case and has two components of slighly different frequency. This leads to two series of hot bands which overlap and, especially at the high frequency side of v_1, smear out the structure. In Fig. 10 taken from Thomas [79] the first six hot band origins are marked. The interpretation of the fine structure leads to 145 and 170 cm^{-1} for the two components of v_β while v_σ can be estimated to be 180 cm^{-1} and the enthalpy of association $\Delta H^0 = -26$ kJ mol^{-1}.

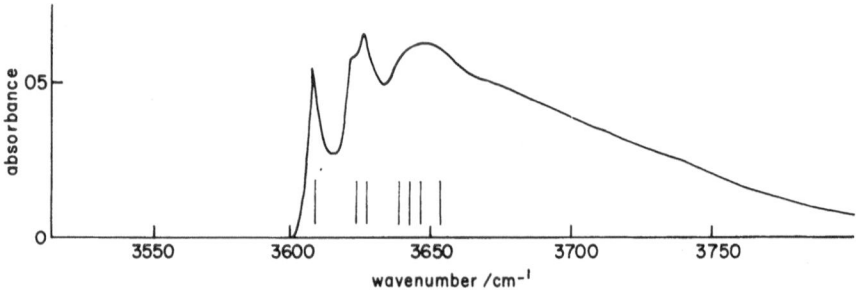

Fig. 10. Infrared spectrum of a water-hydrogen fluoride mixture in the region of the HF stretching vibration. The straight lines indicate the first six hot band origins. From R. K. Thomas, Proc. Roy. Soc. London *A344*, 579 (1975). Reproduced by permission from the Royal Society

The higher bridge deformation vibration is split into two at about 696 and 666 cm^{-1}. It also has rotational fine structure. The v_β bands are at 170 and 145 cm^{-1}. The individual bands are perpendicular bands with strong Q branches whose spacing is altered by Coriolis interaction between the two components of v_β. An interesting treatment of the latter is given in [79].

Curiously, among the internal vibrations of water only the bending vibration v_2 was observed at 1600 cm^{-1}, the stretching vibrations were absent. This is partly due to overlap with the band of free water but their weakness would require further explanation.

Some anharmonic coupling constants were determined: $X_{1b} = 15 \pm 1$ and 18 ± 1 for the two components (696 and 666 cm^{-1}); the average of $X_{b\beta} = 26 \pm 10$ cm^{-1} for the interaction of the two bridge deformation vibrations. These are relatively large values.

It appears clearly that higher resolution IR spectra of a few H-bonded systems in the gas phase increased our knowledge significantly. Rotational fine structure appears as the major cause of the broadening of IR bands affected by H-bond formation in the gas phase. In particular, the structure of the X-H stretching band (v_1) is explained by the presence of satellites due to sum and difference bands of v_1 and the bridge stretching vibration (v_σ) as well as the low frequency bridge bending vibration (v_β) and possible other low frequency modes. The recognition that the observed members of the fine structure are hot bands of v_β containing P-R branches with a band head in the P branch is essential. The importance of anharmonic coupling is underscored by the appearance of the combination bands. Furthermore while the breadth of association bands *is* due to interaction between a high frequency mode and a low frequency bridge vibration, both bridge stretching (v_σ) and bridge bending (v_β) are instrumental in shaping the broad v_1 band.

5 Information Derived From Pure Rotational Spectra

Due to high resolution and large dispersion microwave pure rotational spectra give, in favorable cases, some of the most precise information about molecular structure. Geometrical data can be derived from the rotational constants, dissociation energies from intensity measurements. From the point of view of this review the vibrational satellites, i.e. rotational lines generated by molecules in vibrationally excited states are of special interest. The relative intensities of these vibrational satellites lead to vibrational spacings and possibly to harmonic force constants. Other quantities like vibration-rotation interaction constants and centrifugal distortion constants can give information on both force constants and anharmonicities. Dipole moments can be evaluated from the Stark effect on the rotational spectra. In this way microwave spectra give access to both the potential function and the electronic charge distribution. Cases most favorable for such studies are small molecular species which are either linear or are symmetric tops. Fortunately there are several such systems among the best known gas phase H-bond systems. The purely rotational problems which are involved with these systems are beyond the scope of this review. An attempt is made however, to extract from the available publications all the vibrational information which they contain.

The first full-scale microwave investigation of a H-bonded complex in the gas phase was carried out by Costain and Srivastava [66,67]. These researchers were interested in carboxylic acid dimers. Since for reasons of symmetry cyclic dimers of carboxylic acids have, in general, no dipole moment and consequently no pure rotational spectrum, heterodimers had to be chosen. Costain and Srivastava selected the CF_3COOH-$HCOOH$, CF_3COOH-CH_3COOH and the CH_3COOH-CH_2FCOOH dimers with their deuterated analogues.

The spectrum of CF_3COOH-$HCOOH$ consists of broad lines at intervals of 1154 MHz (1 hertz = $3.34 \cdot 10^{-9}$ cm^{-1}, 1 megahertz = $3.34 \cdot 10^{-3}$ cm^{-1}, 1 gigahertz = 3.34 cm^{-1}) located between 3462 and 21938.5 MHz. From these the rotational constants were evaluated. These lead to an O ... O distance of 2.69 $\pm$ 0.02 A for CF_3COOH-$HCOOH$. Furthermore, it was found that for CF_3COOD-$DCOOD$ this distance is by 0.011 A longer (and weaker). This illustrates the Ubbelohde effect [82,83] which is the consequence of the lesser amplitude and anharmonicity of the O-D vibration. The enthalpy of formation is -66.0 kJ/mol, i.e. 33.0 for each of the two H-bonds (4.18 kJ/mol = 1 kcal/mol). The results were similar for the CF_3COOH-CH_3COOH complex. (R_H = 2.67 A, R_D-R_H = 0.012 A). From R_D-R_H it can be computed that when the OH vibration is excited to v = 1 the O ... O distance is diminished by 0.073 A demonstrating the Sheppard effect. The results also showed that the potential function is a double potential with a barrier of at least 6000 cm^{-1} for CH_3COOH-$HCOOH$ and 5000 cm^{-1} for CF_3COOH-CH_3COOH. This implies rapid tunneling of the proton or deuteron, faster than rotation. This in turn implies an inversion doubling which could contribute to the breadth of the v(OH) vibrational band. As mentioned earlier, a vibrational satellite was found whose intensity indicated a vibrational frequency of 180 cm^{-1} which is readily assigned to one of the ν_σ bridge stretching vibrations.

Bellott and Wilson [72] studied H-bonded dimers formed by trifluoracetic acid with various other carboxylic acids and amides.

In a series of papers Millen and his Coworkers measured and analysed the microwave spectra of the HCN.HF, CH$_3$CN.HF, (CH$_3$)$_3$CCN.HF and H$_2$O.HF complexes. While their full papers were published in 1980, several of their preliminary publications appeared from 1975 to 1978 [84–90].

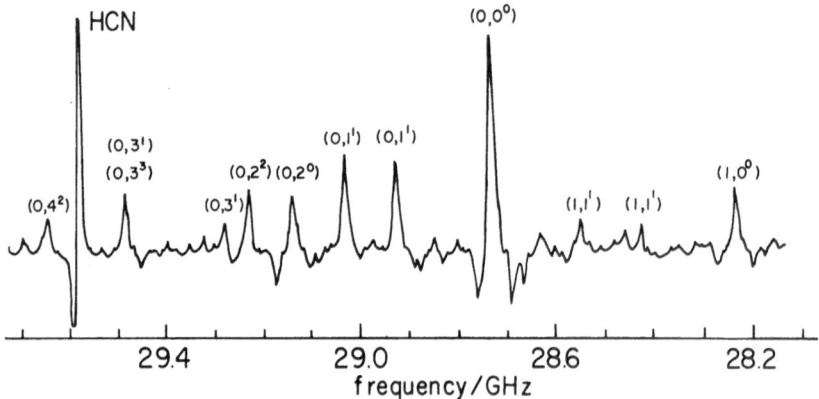

Fig. 11. The N = 4 ← 3 transition of the HCN.HF complex showing the vibrational ground state (0, 0°) and vibrational satellites (ν_σ, $\nu_\beta^{|l|}$). From A. C. Legon, D. J. Millen and S. C. Rogers, Proc. Roy. Soc. London *A370*, 213 (1980). Reproduced by permission from the Royal Society

The HCN.HF system, of fundamental importance was treated by Legon, Millen and Rogers [86–88]. The spectrum gives evidence of a linear structure. (1 type doubling.) Groups of lines were found at 14.4, 28.8 and 36.0 GHz which are readily assigned to J = 2←1, 4←3 and 5←4 transitions where J is the rotational quantum number. (The 1←0 line would be at 7.2) Fig. 11 taken from [88] shows the 4←3 transition. The strongest band represents the 4←3 transition in the vibrational ground state (0, 0). (The upper index refers to the quantum number of the vibrational angular momentum for degenerate vibrations in this linear system.) It is at 28 727.7 MHz, while the 2←1 frequency is found at 14 364.3 and the 5←4 frequency is at 35 908.35 MHz. From these the rotational and centrifugal distortion constants, B$_0$ and D$_0$, can be evaluated through the expression

$$\nu = 2B_0(J + 1) - 4D_0(J + 1)^2 .$$

The other lines in Fig. 11 are vibrational satellites. Their intensities depend on the Boltzmann factor and therefore on the frequencies of the vibrations that are involved. Knowing the temperature these frequencies can be computed from the observed intensities. The vibrations which stand a good chance of causing satellites are, of course, the ones of low frequency. Table 9 is a part of Legon, Millen and

Rogers' Table 1 [88] who assigned all the satellites to v_σ, v_β and their various overtones and combinations. From the intensities they obtained $v_\sigma = 197 \pm 15$ cm^{-1} and $v_\beta = 91 \pm 20$ cm^{-1} in fair agreement with their values obtained by Thomas [78] from the vibrational spectrum. They did not observe satellites with the other bridge bending mode, 555 cm^{-1}, presumably because of its low Boltzmann factor.

Table 9. Observed rotational transition frequencies in the ground state and vibrationally excited states of H^{12}C^{14}N ... H^{19}F. From A. C. Legon, D. J. Millen and S. C. Rogers, Proc. Roy. Soc. London A370, 213 (1980). Reproduced by permission from the Royal Society.

vibrational assignment		frequencies/MHz
v_σ	$v_\beta^{|l|}$	$N = 4 \leftarrow 3$
1	0	28232.85 $\pm$ 0.5
1	1^1	23426.2 $\pm$ 1
1	1^1	28549.7 $\pm$ 1
0	0	28727.7 $\pm$ 0.3
0	1^1	28925.3 $\pm$ 0.3
0	1^1	29028.7 $\pm$ 0.5
0	2^0	29136.1 $\pm$ 0.6
0	2^2	29228.2 $\pm$ 0.5
0	3^1	29278.5 $\pm$ 0.5
0	3^1	29483.3 $\pm$ 0.5
0	3^3	29483.3 $\pm$ 0.5
0	4^0	—
0	4^2	29645.8 $\pm$ 0.7

In addition to spectroscopic constants Legon et al. [88] determined geometrical parameters, dipole moments, dissociation energies and force constants for HCN.HF and a number of its isotopically substituted analogs.

Using the best available values for the internuclear distances they obtained 2.796 A for the N ... F distance. The result was the same for HCN.HF and for HCN.DF. These H-bonds are, of course, weak and this does not contradict the existence of the Ubbelohde effect for stronger H-bonds. The dissociation energies were obtained by a method previously described by Legon, Millen, Mjöberg and Rogers [87]. The zero-point dissociation energy D_0 from a H-bonded system can be obtained from absolute intensity measurements of rotational lines in the equilibrium mixture; the equilibrium dissociation energy D_e can also be computed if the vibrational frequencies are known. They obtained 18.9 $\pm$ 1.1 kJ mol^{-1} for D_0 and 26.1 $\pm$ 1.6 for D_e using the harmonic oscillator approximation.

The dipole moments can be obtained from the Stark effect of the rotational transitions. The mean value was $\mu = 5.612 \pm 0.01$ D, an enhancement of 0.80 D over the vector sum of the monomer values.

In the second paper of the series Bevan, Legon, Millen and Rogers [89] presented the microwave spectrum of CH$_3$CN.HF which showed that the complex is a symmetric top. In the third paper Georgiou, Legon and Millen [90] examined

(CH$_3$)$_3$CCN.HF. The C-C≡N ... HF fragment is linear. As for HCN.HF the Authors obtained a valuable set of spectroscopic constants as well as information on the structure of the complex. We summarize some of the results they obtained in Table 10.

Table 10. N ... F distances and bridge frequencies. Data from A. S. Georgiou, A. C. Legon and D. J. Millen [86].

Complex	N ... F distance	v_σ	v_β
HCN.HF	2.796 A	197 cm^{-1}	91 cm^{-1}
CH$_3$CN-HF	2.759	181	45
(CH$_3$)$_3$CC≡N.HF	2.725		55

For all three systems the HF and DF complex had the same N ... F distance. The more interested Reader will certainly like to consult papers [89] and [90] for the complete set of force constants obtained for the first time for H-bond systems, including the H-bond stretching and bending constants.

In an earlier paper Bevan, Legon, Millen and Rogers [84] presented the microwave spectrum of the water-hydrogen fluoride complex. For the O ... F distance they obtained 2.68 ± 0.01 A and for the dipole moment component along the symmetry axis 3.82 ± 0.02 D. For v_σ they gave, as a preliminary value, 198 cm^{-1} and for the out-of-plane and in-plane bridge vibrations 94 and 180 cm^{-1}, respectively. A full paper has been announced.

This section has been kept purposely short and sketchy since the writer has knowledge of a forthcoming extensive review by Professor D. J. Millen. It should be pointed out, however, that the results mentioned in this Section constitute the most complete and most quantitative results on H-bond systems that had been obtained up to the present time.

6 Concluding Remarks

There is little doubt that our understanding of the nature of the hydrogen bond owes a great deal to attempts of explaining the great breadth and complicated shape of the infrared X-H stretching band. Unfortunately solvent and crystallin environ- ment rendered the access to the individual H-bond difficult. Solid and solution spectra were an arguable check on the many theoretical schemes that were proposed. Spectra of H-bonded species in the gas phase were for a long time badly lacking. Starting 1965 such spectra were forthcoming, however. The vibrational work of Millen's group was followed by Thomas' work on the rotational fine structure of the infrared bands and recently the pure rotational work by Millen's group led to accurate information on several H-bonded systems. Ether.HF, ether.HCl and cyanide-HF systems played a prominent role in this evolution.

Briefly summarizing our present knowledge we can make the following statements. The great breadth of the IR X-H stretching bands is due to combination bands of the

$(\nu_1 \pm \nu_\sigma)$ and the $(\nu_1 \pm \nu_\beta)$ types which accompany ν_1 and under low resolution convolute with it. That in addition to the bridge stretching modes we have to invoke the bridge deformation modes has been revealed mainly by the spectra of deuterium-bonded systems. Among the combinations that contribute to ν_1's breadth hot bands of the $(\nu_1 + n'\nu_\beta - n''\nu_\beta)$ type are most important. The individual bands when they are resolved exhibit a band head in the P branch in this way demonstrating the existence of the Sheppard effect. Anharmonic coupling plays an essential role in making it possible for all these combination bands to appear. For stronger H-bonds several other combinations due to internal modes appear in the IR spectrum. These are likely to borrow intensity through Fermi resonance from the main X-H stretching band in most cases. This is just another manifestation of the increased anharmonicity, characteristic of H-bond systems and it in no way contredicts the involvement of the interaction between the high frequency X-H vibrations and the low frequency bridge stretching and deformation motions. Both mechanical and electrical anharmonicity are needed in order to explain the observed intensities.

While much work remains to be done on individual systems, the writer believes that these phenomena are now essentially understood.

The transition form the gas phase to conditions prevailing in solution and in solids can be made through the application of relaxation theories as has been shown by Bratos [91]. Gradually rotational fine structure is replaced by broadening due to energy dissipation involving the random solvent environment or lattice motions.

All this applies to weak and medium strong H-bonds like those encountered for alcohols and many other systems up to carboxylic acid dimers or about 32–42 kJ/mol. (8 or 10 kcal/mol.) Unfortunately vibrational spectra of systems with very strong H-bonds could, with a few exceptions, only be measured in condensed phases. Factors that come in when such systems are examined are potential surfaces with two minima with, in certain cases, the possibility of tunnelling, or flat single minima. Most of these systems are likely to be so anharmonic that second order perturbation theory breaks down and the concept of normal vibrations becomes itself questionable. Many such systems are highly polarizable and are strongly influenced by the environment yielding extremely broad bands [92]. Bratos and Ratajczak [93] has shown that even such systems can be handled by relaxation theories.

It would be most desirable to have IR spectra of very strongly hydrogen bonded systems in the gas phase. Such systems has been produced in the gas phase through advanced techniques elaborated by Kebarle [94–96] (high-pressure mass spectrometry), Beauchamp (ion cyclotron resonance [97]) and Bohme (flowing afterglow technique [98] and their respective groups.

These are generally ionic species involving net charges as well as H-bonds. Interesting work in this direction has been reported by Schwarz [99] and by Gerasimov, Kulbida, Tokhadze and Schreiber [100]. It can be hoped that through another step forward in experimental techniques IR spectra of such systems will become available in a not too distant future. The writer believes that this would be the logical next stage in the evolution of our knowledge on H-bonds. It may be said, however, as per now, that there is no more mystery about the nature of the hydrogen bond.

Camille Sandorfy

7 References

1. The Hydrogen Bond, Vols I–III. Eds, Schuster, P., Zundel, G. and Sandorfy, C. North-Holland, Amsterdam, 1976
2. Allerhand, A., Schleyer, P. R.: J. Am. Chem. Soc., 85, 371, 1233 (1963)
3. Foldes, A. and Sandorfy, C.: Can. J. Chem. 49, 505 (1971)
4. Wagner, E. L. and Hornig, D. F.: J. Chem. Phys., 18, 296, 305 (1950)
5. Plumb, R. C. and Hornig, D. F.: J. Chem. Phys. 23, 947 (1955)
6. Waldron, R. D.: J. Chem. Phys. 21, 734 (1953)
7. Cabana, A. and Sandorfy, C.: Spectrochim. Acta 18, 843 (1962)
8. Bratos, S., Hadži, D. and Sheppard, N.: Spectrochim. Acta 8, 249 (1956)
9. Bernard-Houplain, M. C. and C. Sandorfy, C.: J. Chem. Phys. 56, 3412 (1972)
10. Bicca de Alencastro, R. and Sandorfy, C.: Can. J. Chem. 50, 3594 (1972)
11. Herzberg, G.: Molecular Spectra and Molecular Structure. Vol. 2, Ed. D. Van Nostrand, New York, 1945
12. Sandorfy, C.: Chapter 13 in [1]
13. Sandorfy, C.: in Infrared and Raman Spectroscopy of Biological Molecules. Ed. Th. Theophanides. D. Reidel, Dordrecht, Holland, 1979, p. 305
14. Stepanov, B. I.: Zhur. Fiz. Khim. 19, 507 (1945)
15. Stepanov, B. I.: Zhur. Fiz. Khim. 20, 408 (1946)
16. Sheppard, N.: in Hydrogen Bonding, Ed. D. Hadži. Pergamon Press, London, (1959), p. 85
17. Robertson, G. N.: Phil. Trans. Roy. Soc. London, 286, 25 (1977)
18. Herzberg, G.: Molecular Sapectra and Molecular Structure. Vol. 1. Ed. D. Van Nostrand, New York, 1950
19. Bertie, J. E. and Millen, D. J.: J. Chem. Soc. 497 (1965)
20. Arnold, J. and Millen, D. J.: J. Chem. Soc. 503 (1965)
21. Arnold, J. and Millen, D. J.: J. Chem. Soc. 510 (1965)
22. Bertie, J. E. and Millen, D. J.: J. Chem. Soc. 514 (1965)
23. Millen, D. J. and Samsonov, O. A.: J. Chem. Soc. 3085 (1965)
24. Millen, D. J. and Zabicky, J.: J. Chem. Soc. 3080 (1965)
25. Arnold, J., Bertie, J. E. and Millen, D. J.: Proc. Chem. Soc. 121 (1961)
26. Belozerskaya, L. P. and Shchepkin, D. N.: Opt. Spectrosc. Molec. Spectrosc. Suppl. 11, 146 (1966)
27. Lassègues, J. C. and Huong, P. V.: Chem. Phys. Lett. 17, 444 (1972)
28. Bertie, J. E. and Falk, M. V.: Can. J. Chem. 51, 1713 (1973)
29. Desbat, B. and Lassègues, J. C.: J. Chem. Phys. 70, 1824 (1972)
30. Lautié, A. and Novak, A.: J. Chem. Phys. 56, 2479 (1972)
31. Bernard-Houplain, M. C. and Sandorfy, C.: Chem. Phys. Lett. 27, 154 (1974)
32. Di Paolo, T., Bourdéron, C. and Sandorfy, C.: Can. J. Chem. 50, 3161 (1972)
33. Luck, W. A. P. and Ditter, W.: J. Mol. Struct. 1, 261 (1967–1968)
34. Maréchal, E. and Bratos, S.: J. Chem. Phys. 68, 1825 (1978)
35. Bouteiller, Y. and Maréchal, E.: Mol. Phys. 32, 277 (1976)
36. Bouteiller, Y. and Guissani, Y.: Mol. Phys. 38, 617 (1979)
37. Witkowski, A. and Maréchal, Y.: J. Chem. Phys. 48, 3697 (1968)
38. Thomas, R. K.: Proc. Roy. Soc. London A 325, 133 (1971)
39. Le Calvé, J., Grange, P. and Lascombe, J.: Compt. Rend. Acad. Sci. Paris 260, 6065 (1965)
40. Couzi, M., Le Calvé, J., Huong, P. V. and Lascombe, J.: J. Mol. Struct. 5, 363 (1970)
41. Bevan, J. W., Martineau, B. and Sandorfy, C.: Can. J. Chem. 57, 1341 (1979)
42. Bernstein, H. J., Clague, D., Gilbert, A., Michel, A. J. and Westwood, A.: Symposium of the Canadian Spectroscopy Society. Toronto. 1980
43. Hussein, M. A. and Millen, D. J.: J. Chem. soc. Faraday Trans. II, 70, 685 (1974)
44. Millen, D. J. and Mines, G. W.: J. Chem. Soc. Faraday Trans II, 70, 693 (1974)
45. Hussein, M. A., Millen, D. J. and Mines, G. W.: J. Chem. Soc. Faraday Trans. II, 72, 686 (1976)
46. Hussein, M. A. and Millen, D. J.: J. Chem. Soc. Faraday Trans. II, 72, 693 (1976)
47. Millen, D. J. and Mines, G. W.: J. Chem. Soc. Faraday Trans. II, 73, 369 (1977)

48. Legon, A. C., Millen, D. J. and Schrems, O.: J. Chem. Soc. Faraday Trans. II, *75*, 592 (1979)
49. Tucker, E. E. and Christian, S. D.: J. Am. Chem. Soc. *98*, 6109 (1976)
50. Fild, M., Swiniarski, M. and Holmes, R.: Inorg. Chem. *9*, 839 (1970)
51. Carlson, G. L., Witkowski, R. E. and Fateley, W. G.: Nature *211*, 1289 (1966)
52. Al-Adhami, L. and Millen, D. J.: Nature *211*, 1291 (1966)
53. Barnes, A. J., Hallam, H. E. and Jones, D.: Proc. Roy. Soc. London A *335*, 97 (1973)
54. Barnes, A. J., Hallam, H. E. and Jones, D.: J. Chem. Soc. Faraday Trans. II, *70*, 422 (1974)
55. Reece, I. H. and Werner, R. L.: Spectrochim. Acta *24A*, 1271 (1968)
56. Krueger, P. J. and Mettee, H. D.: Can. J. Chem. *42*, 347 (1964)
57. Krueger, P. J. and Mettee, H. D.: Can. J. Chem. *42*, 326, 340 (1964)
58. Inskeep, R. G., Kelliher, J. M., McMahon, P. E. and Somers, B. G.: J. Chem. Phys. *28*, 1033 (1958)
59. Clague, A. D. H., Govil, G. and Bernstein, H. J.: Can. J. Chem. *47*, 625 (1969)
60. Kivinen, A. and Murto, J.: Suomen Kem *B40*, 6 (1967)
61. Murto, J., Kivinen, A., Korppi-Tommola, J., Viitala, R. and Hyomaki, J.: Acta Chem. Scand. *27*, 107 (1973)
62. Hadži, D. and Sheppard, N.: Proc. Roy. Soc. London *A216*, 274 (1953)
63. Haurie, M. and Novak, A.: J. Chem. Phys. *62*, 137 (1965)
64. Haurie, M. and Novak, A.: J. Chim. Phys. *62*, 146 (1965)
65. Foglizzo, R. and Novak, A.: J. Chim. Phys. *71*, 1322 (1974)
66. Costain, C. C. and Srivastava, G. P.: J. Chem. Phys. *35*, 1903 (1961)
67. Costain, C. C. and Srivastava, G. P.: J. Chem. Phys. *41*, 1620 (1964)
68. Jacobsen, R. J., Mikawa, Y. and Brasch, J. W.: Spectrochim. Acta *23A*, 2199 (1967)
69. Jacobsen, R. J., Brasch, J. W. and Mikawa, Y.: Appl. Spectr. *22*, 641 (1968)
70. Christian, S. D. and Stevens, T. L.: J. Phys. Chem. *76*, 2039 (1972)
71. Pava, B. M. and Stafford, F. E.: J. Phys. Chem. *72*, 4628 (1968)
72. Bellott, E. M. and Wilson, E. B.: Tetrahedron *31*, 2896 (1975)
73. Gilbert, A. S. and Bernstein, H. J.: Can. J. Chem. *52*, 674 (1974)
74. Jones, W. J., Seel, R. M. and Sheppard, N.: Spectrochim. Acta *25A*, 385 (1969)
75. Gerhard, S. L. and Dennison, D. M.: Phys. Rev. *43*, 197 (1933)
76. Thomas, R. K. and Sir. H. Thompson, Proc. Roy. Soc. London, *A316*, 303 (1970)
77. Thomas, R. K.: Proc. Roy. Soc. London, *A322*, 137 (1971)
78. Thomas, R. K.: Proc. Roy. Soc. London, *A325*, 133 (1971)
79. Thomas, R. K.: Proc. Roy. Soc. London, *A344*, 579 (1975)
80. Brown, J. K. and Sheppard, N.: Spectrochim. Acta *23A*, 129 (1967)
81. Benesi, H. A. and Hildebrand, J. H.: J. Am. Chem. Soc. *71*, 2703 (1949)
82. Robertson, J. M. and Ubbelohde, A. R.: Proc. Roy. Soc. London, *A170*, 222, 241, (1939)
83. Gallagher, K. J.: in Hydrogen Bonding. Ed. D. Hadži. Pergamon Press, London, 1959. p. 45
84. Bevan, J. W., Legon, A. C., Millen, D. J. and Rogers, S. C.: J.C.S. Chem. Comm. 341 (1975)
85. Bevan, J. W., Legon, A. C., Millen, D. J. and Rogers, S. C.: J.C.S. Chem. Comm. 130 (1975)
86. Legon, A. C., Millen, D. J. and Rogers, S. C.: Chem. Phys. Lett. *41*, 137 (1976)
87. Legon, A. C., Millen, D. J., Mjöberg, P. J. and Rogers, S. C.: Chem. Phys. Lett. *55*, 157 (1978)
88. Legon, A. L., Millen, D. J. and Rogers, S. C.: Proc. Roy. Soc. London, *A370*, 213 (1980)
89. Bevan, J. W., Legon, A. C., Millen, D. J. and Rogers, S. C.: Proc. Roy. Soc. London, *A370*, 239 (1980)
90. Georgiou, A. S., Legon, A. C. and Millen, D. J.: Proc. Roy. Soc. London *A370*, 257 (1980)
91. Bratos, S.: J. Chem. Phys. *63*, 3499 (1975)
92. Zundel, G.: in The Hydrogen Bond. Eds. Schuster, P., Zundel, G. and Sandorfy, C. Vol. 2, p. 683. North-Holland, Amsterdam, 1976
93. Bratos, S. and Ratajczak, H.: in press.
94. Kebarle, P.: Ann. Rev. Phys. Chem. *28*, 445 (1977)
95. Davidson, W. R., Sunner, J. and Kebarle, P.: J. Am. Chem. Soc. *101*, 1675 (1979)

Camille Sandorfy

96. Lau, Y. K., Saluja, P. P. S. and Kebarle, P.: J. Am. Chem. Soc. *102*, 7429 (1980)
97. Henderson, W. G., Taagepera, D., Holtz, D., McIver, R. T., Beauchamp, J. L. and Taft, R. W.: J. Am. Chem. Soc. *94*, 4728 (1972)
98. Bohme, D. K., Hemsworth, R. S., Rundle, H. W., Schiff, H. I.: J. Chem. Phys. *58*, 3504 (1973)
99. Schwarz, H. A.: J. Chem. Phys. *72*, 294 (1980)
100. Gerasimov, I. V., Kulbida, A. I., Tokhadze, K. G. and Schreiber, V. M.: Zh. Prikl. Spectr. *32*, 1066 (1980) (p. 629 in the translated Journal)
101. Maréchal, Y.: Chapter 8 in "Molecular Interactions". Edited by H. Ratajczak and W. J. Orville-Thomas. Wiley and Sons, 1980

Microwave and Radiofrequency Spectra
of Hydrogen Bonded Complexes in the Vapor Phase

Thomas R. Dyke

Department of Chemistry University of Oregon Eugene, Oregon 97403, USA

Table of Contents

A Introduction

An enormous amount of research has been directed at the phenomenon of hydrogen bonding [1-3]. This effort has been generated by the importance of hydrogen bonding phenomena in nature. The structure and properties of condensed phases are strongly dependent on weak interactions such as hydrogen bonds and van der Waals bonds. The structure and solvation properties of macromolecules in biological systems are greatly influenced by hydrogen bonding [4]. Atmospheric processes such as the transmission of electromagnetic radiation can be affected by hydrogen bonded molecules [5,6].

The work which is reviewed here provides accurate structural data from microwave and radiofrequency spectroscopy of relatively small molecule, hydrogen bonded complexes. Its role has been to provide information concerning the stereochemistry and electronic properties — electric dipole moments and nuclear hyperfine interactions — characteristic of hydrogen bonds. The experiments are done on gas phase samples, often in molecular beams, which eliminates environmental perturbations of the hydrogen bonds. In addition, the small molecules used are amenable to *ab initio* calculations [7-9] and thus the results are extremely useful as criteria for the accuracy of these calculations. Finally, the results are useful to construct models of more complex systems in chemistry and biology involving hydrogen bonds [4].

The first experiments in this area were low resolution microwave spectroscopy of carboxylic acid complexes done by Costain and Srivastava [10] in the 60's. The discovery that the $(HF)_2$ rotational-tunneling spectrum was conveniently studied by molecular beam electric resonance spectroscopy (Dyke, Howard and Klemperer [11]) led to fruitful studies of a large number of weak (eg. $Ar \cdot HCl$ [12]) and "normal" [eg. $(HF)_2$ [11], $(H_2O)_2$ [13]] hydrogen bonded species. Conventional, high-resolution microwave spectroscopy has also been found to be useful in the study of normal hydrogen bonded complexes by Millen and co-workers [14]. More recently, Flygare's group at Illinois developed and elegant pulsed, molecular beam-microwave Fourier transform technique [15] which they have applied to a large number of hydrogen bonded molecules.

In this review, the experimental techniques in use will be briefly described and then the results of these studies will be presented in moderate detail. At present, spectroscopic results have been obtained only for bimolecular complexes, although some qualitative structural information for larger clusters has been gained by molecular beam, electric deflection techniques. Therefore, the discussion will concern bimolecular complexes. For convenience, these molecules will frequently be referred to as "dimers", whether or not the constituent parts are identical. In addition, since the experiments have illustrated that the constituent molecules essentially retain their identity for these weakly bonded (< 10 kcal/mole) complexes, they will often simply be referred to as the monomers in the complex.

B Experimental Methods

I Microwave Absorption Spectroscopy

The earliest microwave work on hydrogen bonded complexes was the low resolution absorption spectroscopy of carboxylic acid dimers by Costain and Srivastava [10].

More recently, these techniques were extended to several carboxylic acid dimers and mixed carboxylic acid-amide complexes [16].

The carboxylic acid dimers are quite heavy, with rotational constants typically around 1 GHz, and the microwave absorption experiments are conducted at "high" temperatures of 200–300 K. The resulting large number of rotation-vibration states populated, coupled with low dimer number densities, on the order of 5×10^{14} molecules/cm³, makes complete resolution of the rotational spectrum not feasible. However, virtually all dimers are prolate rotors with only moderate asymmetry. Thus, $\Delta J = 1$ transitions (a-type) with the same initial and final quantum numbers, but otherwise of different asymmetric rotor state or different vibrational state, will have the same frequency within about 50 MHz for moderate J values; e.g. for $J < 5$ and for transition frequencies less than 50 GHz. At this level of resolution, isotope shifts are not discernible, and the resulting spectra (Fig. 1) yield one rotational constant, $(B + C)/2$, with an accuracy of about 0.5 %.

From $(B + C)/2$ and known monomer geometries, the distance between the monomers can be calculated, and by inference, the hydrogen bond length. In addition, the electric dipole moment a-component can be measured from line-broadening

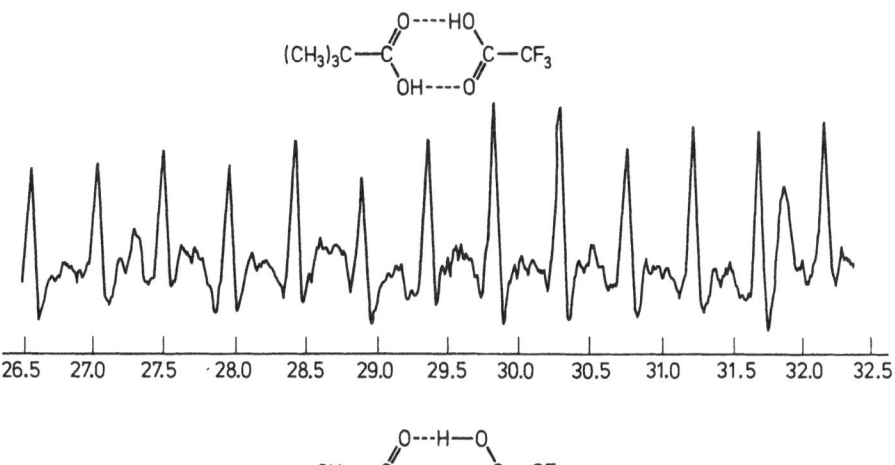

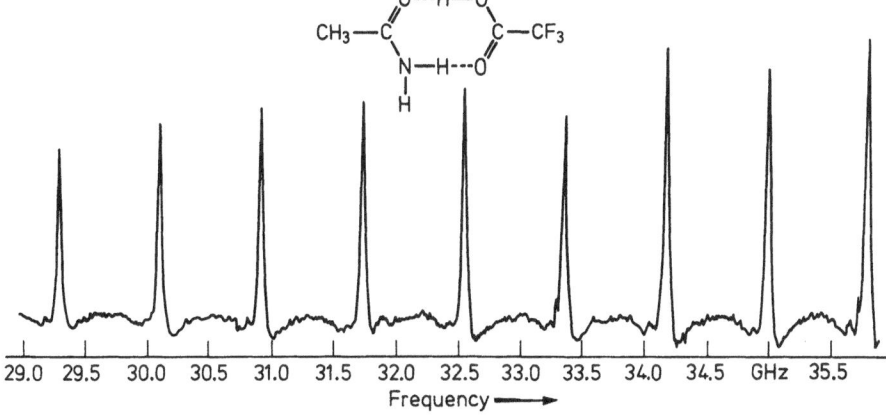

Fig. 1. Low resolution microwave spectra of carboxylic acid dimers. (Reproduced from Ref. [16].)

caused by the Stark effect. Stark *shift* measurements are precluded by the blend of different asymmetric rotor states in each transition, whose Stark effects do not "pile up" nicely, even in the symmetric rotor limit.

More recently, Millen and co-workers [14, 17-25] have examined a number of small molecule dimers, primarily with HF as one constituent, using high resolution microwave absorption spectroscopy [26]. As in the above experiments, conventional absorption techniques are used. By examining dimers with small molecule monomers, the sensitivity is adequate to completely resolve rotational spectra. Thus isotopic substitution methods and Stark shift measurements are possible and accurate dimer geometries and electric dipole moments can be calculated.

One particularly nice feature of these experiments is that intensities of resolved vibrational satellites can be used to determine vibrational frequencies. A dimer with non-linear monomers has six low-frequency (< 600 cm^{-1}) *intermolecular* modes of vibration. At 200–300 K, the corresponding excited vibrational states may have substantial populations and be observable. Although it can be difficult to assign the resulting vibrational satellites [27], particularly if the molecule is undergoing large-amplitude tunneling motions, relative intensity measurements can supply informative vibrational frequencies. In a similar manner, intensity measurements for the monomers and dimers can be used to calculate number densities and hence equilibrium constants for dimer formation [18]. These constants can then be used to estimate the hydrogen bond energy, which is of great interest.

II Molecular Beam Electric Resonance Spectroscopy

Molecular beam techniques have been particularly useful in studying weakly bound complexes, and a large number of studies concerning hydrogen bonded and van der Waals bonded complexes have been carried out using the electric resonance technique [11-13, 31-39]. The first of these studies involved molecular beam electric resonance spectroscopy of $(HF)_2$ [11]. By allowing the monomers of interest to expand in a free-jet nozzle source, molecular beams of dimers in reasonable intensity at low temperatures were produced. Since the properties of such expensions are now well-known [28], it is sufficient to point out that for spectroscopy, it has been found that the highest signal-to-noise ratios are achieved by mixing a few per cent of the monomers of interest with an inert gas carrier. Since the inert gases have no rotation-vibration energy and form complexes with difficulty, such mixtures give extremely low rotational temperatures, less than 10 K, which is highly advantageous for high resolution spectroscopy. Typical dimer beam fluxes in these experiments are 10 [15] dimers/steradian-sec. Although higher dimer fluxes can be achieved by using "neat" gases, this is usually more than offset by the increased internal temperatures of the beam molecules.

Molecular beam electric resonance spectroscopy has been reviewed a number of times [29, 30]. In brief, inhomogenous electric fields are employed to deflect molecular beams via the Stark effect. If the Stark energy is $C_2\mu^2E^2$ and the electric field gradient $\vec{\nabla E}$, then the deflection force is simply $\vec{F} = C_2\mu^2E\vec{\nabla E}$. Since the Stark effect depends strongly on the rotational state of the molecule (through C_2), electric dipole transitions are detected by altering the trajectories of molecules which have undergone a rotational transition. By allowing molecules on only certain trajectories to reach the detector (electron-impact ionizer, mass spectrometer), a plot of the number of molecules detected vs. frequency of the radiation field reveals the spectrum.

The principle advantages of this method include elimination of collision broadening effects and lessening of Doppler broadening effects. Spectral line-widths are therefore 1–10 kHz, primarily characteristic of the transit time of the molecules through the radiation field. This is to be compared with the typically 1 MHz wide lines for microwave absorption work — higher than is typical because the high pressure of polar molecules necessary to generate dimers results in large pressure broadening effects. The resolution of the electric resonance method is adequate to obtain precision electric dipole moments and nuclear hyperfine interactions, which give useful structural data (Fig. 2). The extreme low temperature of the molecular beams

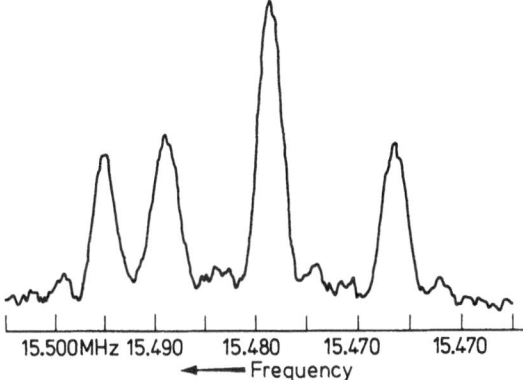

Fig. 2. $J = 1$, $M_J = 0 \to 1$ molecular beam electric resonance for $(H_2O)_2$ at an electric field strength of 599.50 V/cm. The triplet nuclear spin state gives rise to the four spin-spin components of this transition. (Reproduced from Ref. [13 c].)

is also an advantage in simplifying the spectra, but has the disadvantage of eliminating excited vibration-rotation state data which could be useful in determining the potential energy surface for these molecules. Similarly, thermodynamic quantities such as bond energies cannot be estimated because of the non-equilibrium nature of the molecular beam sources.

III Pulsed Molecular Beam, Fourier Transform Microwave Spectroscopy

In the last few years, an elegant pulsed molecular beam technique [15] coupled to Fourier transform microwave spectroscopy [40-42] has been developed by the late W. H. Flygare and his research group. In these experiments, the molecular beam is directed to a Fabry-Perot cavity where a microwave pulse of a few microseconds duration is used to polarize molecules with rotational transitions within the cavity bandwidth (on the order of 1 MHz). After the microwave pulse is turned off, the molecular coherent emission signal is detected with a superheterodyne method and the frequency spectrum gained by Fourier transforms. Examples [43] of these spectra are shown in Fig. 3.

The linewidths are quite small, roughly 10 kHz, and are caused by a Doppler dephasing as the molecules with a given polarization phase move into a region of space where they would have a different phase. There is in addition a splitting of each line (~25 kHz) caused by Doppler effects. Nonetheless the resolution is nearly as good as in the electric resonance experiments, allowing precision rotational constants and nuclear hyperfine interactions to be measured.

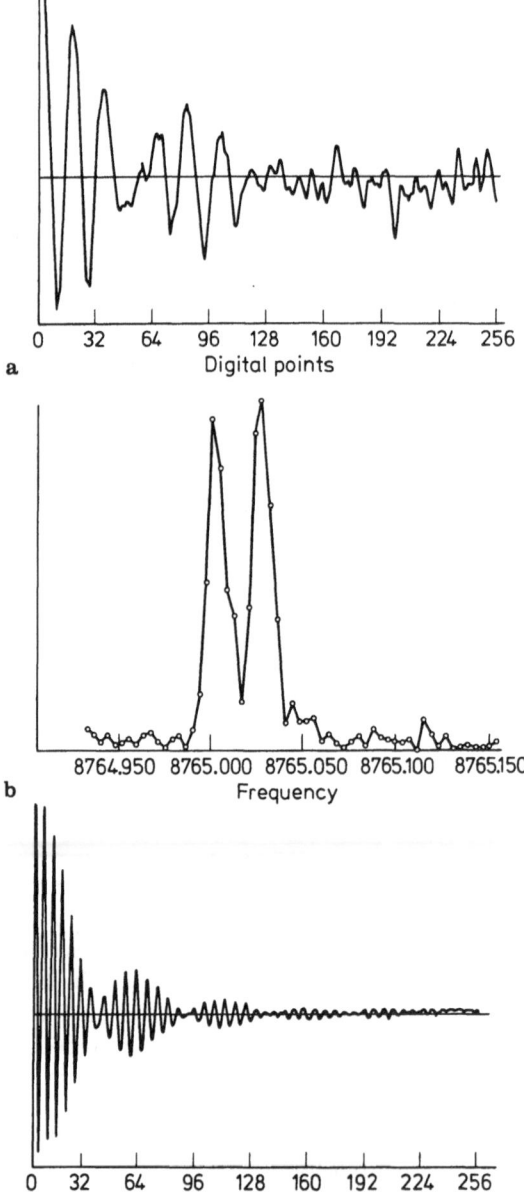

a

b

c

Fig. 3. a) Transient emission signal of $Ar \cdot D^{79}Br$ $J = 3 \rightarrow 4$ transition. b) Power spectrum of part (a). The 23.4 kHz splitting is due to the Doppler effect. c) Transient emission signal from OCS $J = 0 \rightarrow 1$ transition. (Reproduced from Ref. [43].)

The signal-to-noise ratios in these experiments are quite high due to three factors. By working in the time domain, a factor $(\Delta v_p / \Delta v)^{1/2}$ is gained in which Δv_p (~ 1 MHz) is the reciprocal of the microwave pulse time and Δv (~ 10 kHz) is the transition linewidth. Secondly, as in the electric resonance experiments, the rotational temperature of the molecules in the nozzle expansion is very low, only a few degrees Kelvin. This is particularly important in these experiments, since the population inequalities depend on temperature rather than on state selecting fields as in the electric resonance method. Thirdly, by pulsing the molecular beam, detector noise can be limited since the detector need be on only for short periods of time, and yet the full capacity of the pumping system can be utilized. The resulting signal-to-noise ratios are high and a large number of systems have been studied in a short time using this technique [15,43-57].

C Analysis of Spectra

I Rotational Spectra

All of the experimental methods discussed above are used to study the rotational spectra of dimers. The main object is to extract rotational constants for a number of isotopically substituted analogs, and from that data, to calculate the geometry of the dimer. The process is aided by the apparent simplicity of the rotational spectra. Many of the dimers discussed below are either linear molecules or slightly asymmetric rotors in the prolate limit. The rorational energy levels are then given by

$$W = \frac{B+C}{2} J(J+1) + \left(A - \frac{B+C}{2}\right) K^2 + \Delta_{JK} \tag{1}$$

where Δ_{JK} represents splittings caused by asymmetric rotor effects (K-doubling) or vibrational effects (1-doubling). The rotational constants in Eqn. (1) are effective constants which depend on the vibrational and rotational state; e.g. the centrifugal distortion can be explicitly represented:

$$\left(\frac{B+C}{2}\right)_{J,K,v} = \left(\frac{B+C}{2}\right)_{v}^{0} - D_J J(J+1) - D_{JK} K^2 + \dots \tag{2}$$

Since the frequency of $\Delta K = 1$ transitions is often too high to be accessible with the methods discussed above, a-type $\Delta J = 1$ transitions are typically observed, and $(B+C)/2$ constants with the centrifugal distortion parameters determined.

The simplicity of the above picture is misleading in that although the spectroscopy should be straightforward, the non-rigidity of dimers can lead to important effects. For medium strength hydrogen bonds, stretching force constants in the range of 10–20 N/m (1 N/m = 10^{-2} millidyne/Å) and bending force constants of 1–10 J/radian² are observed. For hydride monomers these correspond to r.m.s. vibrational amplitudes of 0.1 Å and 25° for hydrogen bond stretching and bending, respectively, of the dimer.

The most spectacular result of this non-rigidity is that dimers with permutation symmetry high enough can undergo tunneling between isoenergetic conformations of the molecule [58]. The low reduced mass of hydrides and the low barriers to such motions presented by dimers ensure that such complications will occur. Thus both $(HF)_2$ and $(H_2O)_2$ have an a-type rotational spectrum in which the transitions are displaced from Eqn. (1) by $\pm v_t$, where the tunneling doubling is caused by a rapid exchange of donor-acceptor roles for the substituent monomers, and is about 20 GHz in each case [11,59]. The $(H_2O)_2$ spectrum is particularly complex as it also possesses a-type transitions which do not show this doubling. Subtler effects occur for complexes such as $H_2O \cdot HF$ and $H_2S \cdot HF$ in which an internal rotation of 180° around the H_2O or H_2S C_{2v} axis leads to exchange of identical nuclei. In this case the a-type spectra are similar to that of a rigid rotor, but ortho-para type nuclear spin statistics are observed because of the rapid tunneling [22,34].

In addition to tunneling effects, a more general result of the non-rigidity is observed because the molecular constants determined are averages over the vibration-rotation wavefunction. In the extreme case of molecules like $Ne \cdot HCl$ [37,65], constants such as the electric dipole moment components may change by a factor of two upon deuterium substitution. In the less extreme case of medium strength hydrogen bonds (~ 5 kcal/mole), it has nevertheless been possible to determine structures from rotational constant data using rigid-rotor models. Generally, in these cases, the monomer geometries have been assumed to be those of the free monomer, and rotational constant data for various isotopically substituted dimers used to cal-calculate geometries, assuming rigid rotor behavior.

Although the r.m.s. vibrational amplitudes are quite large, the above procedure is reasonable since the vibrational averaging frequently leaves quantities such as rotational constants within a much narrower envelope centered around the equilibrium value than is suggested by the r.m.s. vibrational amplitude. By choosing rotational constants which are relatively intensive to these vibrational averaging effects, hydrogen bond lengths to within a few hundredths of an Angström and angles to within 10° of the equilibrium values can be calculated. However, the situation is not entirely satisfactory, and refined treatments, presumably combining scattering data and vibrational spectra as well, to obtain the full potential energy surface will be important. In the case of the inert gas, hydrogen halide systems, the rigid rotor models fail rather badly. A number of treatments alluded to above have been carried out [61-65] and will be discussed later on in conjunction with the experimental results for those molecules.

II Stark Effects

In addition to moment-of-inertia data, microwave spectra and radiofrequency ($\Delta J = 0$, $\Delta M_J = 1$) spectra can be used to measure Stark effects quite accurately. Electric dipole moments good to about 1% have been calculated from the static, high resolution microwave experiments, and moments with less than 0.01% error are routinely determined with the electric resonance method.

Because of the very prolate nature of many dimers, their Stark effects can often be simply analyzed with the linear rotor expression:

$$W_S = \frac{\mu_a^2 E^2 [J(J+1) - 3M^2]}{2[(B+C)/2] J(J+1)(2J+3)(2J-1)} + C_{JKM} \frac{\mu_\perp^2 E^2}{A} \qquad (3)$$

where μ_a is the a-component of the dimer moment. The second term of Eqn. (3) is a small correction term for b- and c-type moments [13], with $\mu_\perp = [\mu_b^2 + \mu_c^2]^{1/2}$.

The electric dipole moment results provide useful structural information. In the case of very weak binding, as in inert gas, hydrogen halide systems, the μ_a values yield $\langle \cos \gamma \rangle_a$, where γ is the angle between the hydrogen halide axis and the a-principle axis of inertia of the dimer. In more strongly bound cases, large enhancements of the dipole moment relative to the free monomer values are observed, and rather than provide orientation information, a useful measure of the shift in the charge distribution upon formation of the hydrogen bond is obtained.

III Nuclear Hyperfine Interactions

The resolution of the molecular beam experiments is high enough to observe even rather small nuclear hyperfine interactions such as the spin-spin and spin-rotation interactions as well as the larger quadrupole coupling interactions. The largest terms in the Hamiltonian for the hyperfine splittings are given below [66]:

$$H_{hfs} = Q_i \cdot VE_i + \frac{\mu_i \cdot \mu_j - 3(\mu_i \cdot R)(\mu_j \cdot R)/R^2}{R^3} \qquad (4)$$

The quadrupole coupling interaction is represented by the first term of Eqn. (4), in which Q_i is the quadrupole moment of nucleus i and VE_i is the electric field gradient at this nucleus due to the other charges in the molecule. The second term gives the magnetic dipole-dipole interaction of two nuclei with magnetic moments μ_i and μ_j, separated by a distance R.

The pertinent nuclear moments, fields gradients and geometries are known for most small molecules [29] used to form dimers. To a reasonable approximation, these quantities do not change upon formation of a hydrogen bond, and thus inspection of Eqn. (4) shows that these interactions provide data concerning the orientation of the monomers in the complex. This can be seen more clearly by transforming Eqn. (4) to an effective Hamiltonian [66,67] for calculating matrix elements diagonal in the rotational angular momentum operator, J, and the nuclear spin angular momentum operators, I_i:

$$H_{eff} = \langle eqQ \rangle F(I_i, J) + \frac{(\mu_i/I_i)(\mu_j/I_j)}{R^3} \left\langle 3 \frac{R_z^2}{R^2} - 1 \right\rangle f(I_i, I_j, J)$$

$$F(I, J) = \frac{3(I \cdot J)^2 + \frac{3}{2}(I \cdot J) - I(I+1)J(J+1)}{2I(2I-1)(2J-1)(2J+3)} \qquad (5)$$

$$f(I_i, I_j, J) = \frac{[3(I_i \cdot J)(I_j \cdot J) + 3(I_j \cdot J)(I_i \cdot J) - (I_i \cdot I_j)(J \cdot J)]}{J(2J-1)}$$

In Eqn. (5), the angular brackets impley averages over the asymmetric rotor wavefunction as well as the vibrational wavefunction. R_z is the component of R along the space-fixed z-axis. The final step is to relate the coupling constants in Eqn. (5) to those of the monomer. In general, the expressions depend on the complexity of the monomers and on the dimer rotational state observed. For a large number of cases, a linear type dimer in a K=0 rotational state may be assumed, and Eqn. (5) may be expressed as

$$H_{eff} = \langle eqQ \rangle_{monomer} \left\langle \frac{3 \cos^2 \gamma - 1}{2} \right\rangle F(I_i, J)$$
$$+ \left[\frac{(\mu_i/I_i)(\mu_j/I_i)}{R^3} \right]_{monomer} \left\langle \frac{3 \cos^2 \gamma - 1}{2} \right\rangle f(I_i, I_j, J) \qquad (6)$$

Since the information inside the brackets of Eqn. (6) is available from monomer results, analysis of the dimer hyperfine splittings with Eqn. (6) gives $(3 \cos^2 \gamma - 1)/2$, where γ is the angle between the symmetry axis of the monomer in question and the a-principle axis of the dimer. For non-linear monomers and dimers, Eqn. (6) must be modified, but similar information is obtained.

In some cases, the approximation that the monomer coupling constants in Eqn. (6) are unchanged by hydrogen bond formation is not completely valid. One interesting example is that the quadrupole coupling interaction for a freee inert gas atom is zero, but the presence of a perturbing molecule in a complex such as $Kr \cdot HCl$ [56] or $Xe \cdot HF$ [36] can generate a field gradient at the nucleus of the inert gas. The measured quadrupole coupling splittings then give direct information about this perturbation.

D Experimental Results

I Cyclic Dimers of Carboxylic Acids and Amides

Dimers of CF_3COOH with other carboxylic acids and with various amides have been studied by low resolution microwave spectroscopy. CF_3COOH was chosen for these studies because its large dipole moment enhanced the microwave absorption. These molecules form cyclic dimers [68,69] with two hydrogen bonds in an eight-membered ring. The dimers are readily formed in this stable, two hydrogen bond conformation, with $\Delta H \sim 15$ kcal/mole of dimers.

As mentioned in section B. 1., only one rotational constant, $(B + C)$, per molecule is available from the observed a-type spectrum $(\Delta K = 0, \Delta J = 1)$, and the results are given in Table 1. $(B + C)/2$ is essentially an effective diatomic molecule rotational constant, and therefore depends most strongly on the distance between the monomers of the complex. The rotational constants in Table 1 were best fitted with a model in which O ... H—O distances were 2.67 Å and O ... H—N distances were 2.71 Å. These O ... H—O distances are somewhat shorter than the 2.73 Å distance in formic acid dimer and the 2.76 Å distances in acetic acid dimer [68]. However, they are substantially longer than O ... H—O distances found in crystalline carboxylic acids, e.g. 2.58 and 2.61 Å for formic and acetic acids, respectively [1,70,71]. These crystals are composed of long, hydrogen bonded chains rather than cyclic dimers as in the gas phase. Although the change is not as large as for the dimers discussed in the next section, it is possible that a cooperative effect is causing shorter and stronger hydrogen bonds in the crystal. This speculation is particularly interesting in that the cooperative effects must be extended over more than two hydrogen bonds since the dimer itself possesses two such bonds.

Table 1. (B + C) rotational constants from low resolution microwave spectra of complexes of CF_3COOH with various carboxylic acid and amide partners. An estimate of μ_a from Stark broadening measurements is also given. The (B + C) constants were calculated with O ... H—O distances of 2.67 Å and O ... H—N distances of 2.71 Å.

CF_3COOH Partner	Observed[a] (B + C) MHz	Obs.-Calc.[a] (B + C) MHz	μ_a[a] Debye
$HCOOH$[b]	1154.4	+ 0.3	
CH_3COOH[b]	833.8 (2.0)	+ 1.4	2.99 (.50)
CH_2FCOOH[b]	611.4	+ 0.1	
CH_3CH_2COOH	636.9 (3.0)	+ 0.3	
$(CH_3)_2CHCOOH$	533.2 (2.0)	+ 1.0	
$(CH_3)_3CCOOH$	466.0 (2.0)	+ 1.9	2.28 (.50)
$CH_2=CHCOOH$	656.0 (4.0)	+ 3.1	1.30 (.50)
$(CH_3)_2C=CHCOOH$	420.5 (3.0)	−14.2	0.81 (.50)
▷—COOH	539.6 (2.0)	+ 8.9	2.53 (.50)
⬜—COOH	439 (10)	+ 9	
⬠—COOH	382 (13)	− 5	
CH_3CONH_2	817.9 (2.0)	+ 1.4	3.09 (.50)
$CH_3CH_2CONH_2$	624.4 (2.0)	+ 0.7	2.73 (.50)
(pyrrolidinone) =O, N—H	579.1 (2.0)	− 7.4	1.75 (.50)

[a] Ref. 16. [b] First observed by Ref. 10.

Although the structural detail from these experiments is rather limited, intensity measurements can be used to obtain thermodynamic data. Thus $\Delta H = -15.8$ kcal/mole of dimers was found [10] for the formation of $CF_3COOH \cdot HCOOH$, by observing the temperature dependence of the absorption coefficient and using the relation

$$\frac{d(\ln K)}{dT} = \frac{\Delta H}{RT^2} \cdot \qquad (7)$$

II First and Second Row Hydride Dimers

1 Geometries

High resolution microwave techniques have been applied to several dimers involving first and second row hydrides. Since the high resolution spectra provide accurate frequency shifts upon isotopic substitution, extensive rotational constant data can be gathered. Although, in principle, every coordinate in the molecule could be determined if sufficient isotopic species were studied, in most cases monomer geometries were assumed to be those of the free monomers and the spectra analyzed with rigid rotor models. The resulting geometries (Table 2) reflect zero-point

Table 2. Geometries for first and second row hydrides. R is the heavy atom separation and θ_a is the angle made by the line connecting the heavy atoms and the symmetry axis of the proton acceptor monomer.

	R (Å)[a]	θ_a (degrees)[b]	Comments
F—H----F (with H on acceptor F)	2.79	72	linear H-bond
O—H----O (H₂O····OH₂)	2.98	57	linear H-bond; has plane of symmetry[d]
F—H----O (···OH₂)	2.69[c]	46[c]	linear H-bond; has plane of symmetry[d]
O—H----N (H₂O····NH₃)	2.98	[0]	—
Cl—H----F (with H on acceptor F)	3.37	50	linear H-bond
F—Cl----F (with H on acceptor F)	2.766	55	"anti"-H-bond; planar
F—H----S (····SH₂)	3.25	91	linear H-bond; has plane of symmetry[d]

[a] Uncertainties are roughly 0.33 Å.
[b] Uncertainties are roughly 10°.
[c] Calculated from the data of Ref. 22. Ref. 22 has R = 2.66 Å with θ_a fixed at a planar configuration.
[d] The H_2O and H_2S monomer planes are perpendicular to the symmetry planes of the respective dimers.

vibrational averaging for different isotopic species, and this is the main source of uncertainty in the results shown.

In all cases given in Table 2, the hydrogen bonds are linear, within roughly 10° limits imposed by the rigid-rotor analysis. For $(HF)_2$, $(H_2O)_2$, and $NH_3 \cdot H_2O$, the heavy atom separations for the dimers are respectively 0.30 Å, 0.14 Å and 0.14 Å shorter than in the crystalline solids [1,72,73], presumably reflecting weaker hydrogen bonds for the dimers compared to the solids. Such cooperative effects have been predicted [74,75] and can be explained by electron charge transfer from the electron donor to the electron acceptor molecule. Since the donor-acceptor roles are reversed when a second hydrogen bond is formed, the charge transfer enhances the formation of a second or third hydrogen bond.

A second remarkable trend illustrated in Table 2 concerns the orientations of the proton acceptors, which are roughly tetrahedral for first row hydrides, in contrast to the right-angle geometry of $H_2S \cdot HF$. This is analogous to the trends in the monomers — angles of 104.5° for H_2O and 107° for NH_3 compared to 92° for H_2S and 94° for PH_3. This suggests that the hydrogen-bond has substantial covalent character. This is further borne out by the complexes of oxirane [24] and oxetane [23] with HF shown in Fig. 4. In both cases, the structure can be rationalized in terms

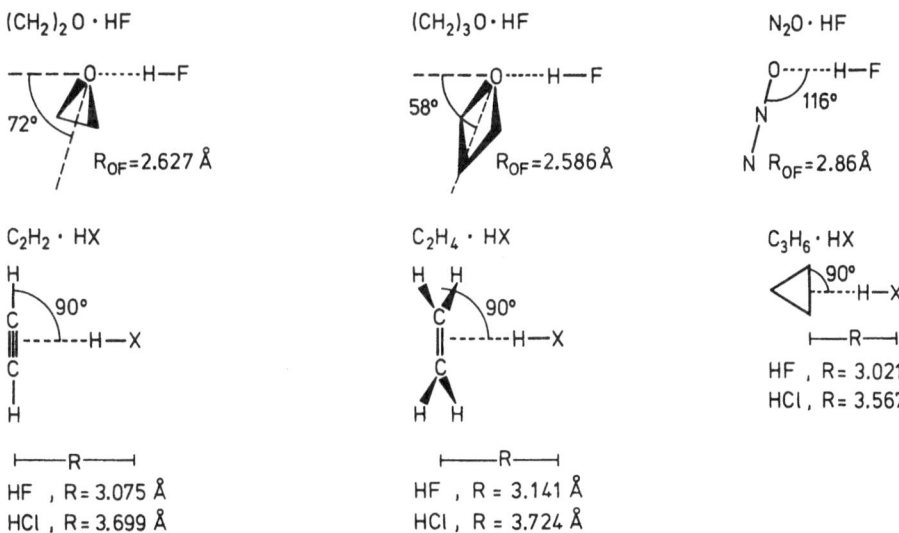

Fig. 4. Structures for non-linear, hydrogen bonded complexes

of a roughly sp^3 hybridized lone pair of electrons on the oxygen oriented along the hydrogen bond axis and donated toward the HF. Given the acute C—O—C oxirane angle of 60° and the less strained C—O—C oxetane angle of 90°, one might expect the lone pairs in oxirane to be further apart than in oxetane or water. The observed angles made by the plane of the electron donor with the hydrogen bond axis are 72° for oxirane, 58° for oxetane, and 57° for H_2O in $(H_2O)_2$, compared to 54.74° for a purely tetrahedral geometry.

These results are in nice agreement with the HOMO-LUMO model discussed by

Klemperer and co-workers. [31] In this model the hydrogen bond is viewed as an electron donor-acceptor complex in which a pair of electrons from the highest occupied molecular orbital of the Lewis base is donated to the lowest unoccupied molecular orbital of the Lewis acid. If the donor electron pair is assumed to have the appropriate hybridization, and the acceptor orbital to be axially symmetric, the above structures can be rationalized as giving maximal overlap between the HOMO and LUMO.

Other structural trends are illustrated by the geometries shown in Table 3 for (essentially) linear hydrogen bonded complexes. A successive shortening of the

Table 3. Geometries for linear hydrogen bonded complexes. $CH_3CN \cdot HF$ and $(CH_3)_3CCN \cdot HF$ are also included in this group. The hydrogen bond lengths are calculated with free monomer bond lengths, assuming a strictly linear geometry.

Complex[a]	Heavy-Atom Separation (Å)	Hydrogen-Bond Length (Å)	Ref.
$HCN \cdot HC'N'$	$R_{NC'} = 3.230$	2.17	[25,27]
$DCN \cdot DC'N'$	$R_{NC'} = 3.232$		
$(CH_3)_3CCN \cdot HF$	$R_{NF} = 2.725$	1.80	[20]
$CH_3CN \cdot HF$	$R_{NF} = 2.759$	1.83	[19]
$CH_3CN \cdot DF$	$R_{NF} = 2.753$		[19]
$HCN \cdot HF$	$R_{NF} = 2.796$	1.87	[18]
$HCN \cdot DF$	$R_{NF} = 2.792$		[18]
$HCN \cdot HCl$	$R_{NCl} = 3.4047$	2.12	[54]
$HCN \cdot DCl$	$R_{NCl} = 3.3993$		
$NCCN \cdot HF$	$R_{NF} = 2.862$	1.94	[47]
$N_2 \cdot HF$	$R_{NF} = 3.082$	2.16	[48]
$OC \cdot HF$	$R_{CF} = 3.047$	2.12	[45]
$OC \cdot DF$	$R_{CF} = 3.036$		[45]
$OC \cdot HCl$	$R_{CCl} = 3.694$	2.41	[44]
$OC \cdot DCl$	$R_{CCl} = 3.684$		[44]
$OC \cdot HBr$	$R_{CBr} = 3.917$	2.50	[46]
$OCO \cdot HF$	$R_{OF} = 2.93$	2.00	[33]
$SCO \cdot HF$	$R_{OF} = 2.96$	2.03	[33]

[a] ^{12}C, ^{14}N and ^{35}Cl isotopic species.

hydrogen bond is observed for the series $RCN \cdot HF$, $R = CN$, H, CH_3 and $C(CH_3)_3$. RCN is the electron donor in these complexes and the shortening correlates with the relative electron releasing character attributed to those groups. Presumably, then, the better electron releasing groups promote stronger and shorter hydrogen bonds. [20,47] Secondly, the increasing hydrogen bond lengths for the series OC ... HX, H=F, Cl, and Br, correlate well with the traditional idea that the relative order of hydrogen bonding ability for the hydrogen halides is HF > HCl > HBr. Finally, the arrangement OC ... HX [76] and SCO ... HF [33] is what would be predicted from the signs of the electric dipole moments; i.e., O^+C^- and S^+CO^-.

For a number of dimers listed in Table 3, a large enough number of isotopic species are available to overdetermine the structural problem, and separate bond lengths are given for pairs of dimers differing only by deuterium substitution. In most cases, the change in hydrogen bond length is less than 0.01 Å, with the deuterated species generally having a shorter heavy atom separation. The small magnitude of the differences are in agreement with what is observed in X-ray diffraction studies of crystalline materials [77], but in those studies, the deuterated species usually have a *longer* heavy atom separation. Since these differences are almost certain to be due to the subtle interplay of monomer and dimer vibrational zero-point effects, it is not surprising that such discrepancies exist. Naively, the dimer results can be rationalized in part by assuming a constant BH distance for a B ... H—A bond and noting that the r.m.s. H—A distance will decrease upon deuteration (0.002 Å for the free monomeric HF and DF [78]).

In Fig. 4, the structures of a number of non-linear complexes are shown. The structure of $N_2O \cdot HF$ can be accounted for by assuming that oxygen donates an sp^2-sp^3 hybridized lone pair to the HF. Similarly, the carboxylic acid structures discussed earlier and preliminary results for the $H_2CO \cdot HF$ complex [79] can be rationalized by an sp^2 hybridized lone pair of electrons from the carbonyl oxygen oriented along the O ... H—O and O ... H—F axes. These results are in sharp contrast to the linear arrangement found for $OCO \cdot HF$ and $SCO \cdot HF$ (Table 3). This is particularly puzzling since CO_2 is isoelectronic with N_2O and shares similar physical properties [33].

Also shown in Fig. 4 are several "non-traditional" complexes of HF and HCl with acetylene, ethylene and cyclopropane. The carbon—carbon bonds for these molecules have relatively high electron densities, and thus the hydrocarbon is acting as an electron donor to the HF or HCl. It is particularly important to note that the strength of the hydrogen bond as reflected in the bond length appears to be comparable to other more traditional HF and HCl hydrogen bonds. This behavior is also reflected in model estimates of stretching force constants (Sect. D.II.4).

2 Electric Dipole Moments

Electric dipole moments components can be calculated from Stark effect measurements [Eqn. (3)], and the results for dimers are collected in Table 4. As shown by Eqn. (3), these experiments give the dipole moment components projected along the principle axes of inertia, but do not directly give the total moment of the molecule. Many of the molecules in Table 4 have A rotational constants which are very large, and thus the Stark effects are dominated by the a-component of the dipole moment in Eqn. (3). For example, μ_a for $H_2S \cdot HF$ is accurately determined to be 2.6239(17)D, but only the combination $\mu_\perp = (\mu_b^2 + \mu_c^2)^{1/2} = 0.97(20)D$ can be found for the remaining components. [34] The total molecular moment is then roughly 2.80(7)D.

In all cases studied, the a-dipole moment component is increased beyond the value calculated from vector addition of the free monomer moments by 0.1–1 Debye. The large dipole moments and polarizabilities of the monomer substituents for the molecules in Table 4 suggest that these enhancements are electrostatically induced:

$$\mu_{IND} = \mu_{IND}^A + \mu_{IND}^B$$
$$= \alpha_A \cdot E_B + \alpha_B \cdot E_A \qquad (8)$$

where α_i is the polarizability of one monomer and E_j is the electric field at that monomer caused by the other substituent of the complex. In many instances, the electric dipole and quadrupole terms of a multipole expansion for E_A and E_B can be

Table 4. Electric dipole moment components and the enhancement, $\Delta\mu_a$, of the dimer μ_a over vector addition of free monomer moments.

Dimer	μ_a (Exp.) Debye	$\Delta\mu_a$ (Calc.) Debye	$\mu_\perp^a$ Debye	Ref.
$(HF)_2$	2.98865(9)	0.60		11)
$HF \cdot DF$	3.0269(12)			11)
$(H_2O)_2$	2.6429(2)	0.46	0.38(10)	13)
$H_2O \cdot HF$	4.07(1)	0.96		22)
$NH_3 \cdot H_2O$	2.972	0.37		80)
$HF \cdot HCl$	2.4095(5)	0.14		31)
$HF \cdot DCl$	2.483(1)			31)
$H_2S \cdot HF$	2.6239(17)	0.78	0.97(20)	34)
$H_2S \cdot DF$	2.669(19)			34)
$HCN \cdot HF$	5.612(10)	0.80		17)
$(CH_2)_2O \cdot HF$	3.85(2)	0.99		24)
$N_2O \cdot HF$	2.069(4)		$\mu_b = 0.69(1)$	32)
$N_2O \cdot DF$	2.125(6)			32)
$CO_2 \cdot HF$	2.2465(4)	0.60		33)
$CO_2 \cdot DF$	2.3024(1)			33)
$SCO \cdot HF$	3.208(2)	0.84		33)

^{a} $\mu_\perp = (\mu_b^2 + \mu_c^2)^{1/2}$

evaluated [82]. Typically the calculation gives an enhancement of a few tenths of a Debye, which is the correct order of magnitude. For example, the dipole and quadrupole contributions to Eqn. (8) for $(H_2O)_2$, give an enhancement of 0.4 Debye, in good agreement with the experiment.

In some cases, the enhacement is larger than is predicted from the electrostatic calculation; e.g., the enhancement for $H_2S \cdot HF$ is 0.3–0.6 D larger than predicted from the dipole and quadrupole term of Eqn. (8). Although part of the discrepancy may lie in the non-convergence of the multipole expansion at these close distances, it is also quite likely that the charge transfer effects implicit in the electron donor-acceptor model mentioned earlier are contributing to the enhancement. A transfer of 1/40 of an electron over a 3 Debye distance generates a 0.36D dipole moment. Analysis of Mulliken populations from *ab initio* calculations [83] indicate that 1/40–1/20 of an electron is a reasonable range for charge transfer in medium strength hydrogen bonds. It is quite likely, then, that both induced moment and charge transfer effects are important contributiors to the observed enhancements.

3 Nuclear Hyperfine Interactions

Nuclear hyperfine splittings in the rotational spectra of dimers have been observed in the molecular beam electric resonance experiments and the Fourier transform microwave experiments. In most cases, the coupling constants are interpreted with the simplified expression given in Eqn. (6) for axially symmetric molecules in the $K=0$ rotational manifold. Thus both the nuclear quadrupole coupling term and the nuclear spin-spin interaction have been used to find $\left\langle \frac{3}{2} \cos^2 \gamma - \frac{1}{2} \right\rangle$, where the brackets imply a vibrational average and where γ is the angle between the symmetry axis of the appropriate monomer and the instantaneous a-axis of the dimer.

The orientations of the monomers for a number of complexes found from hyperfine interactions are given in Table 5 as effective angles γ_A for the proton acceptor and γ_D for the proton donor; i.e.

$$\gamma_A = \arccos\left[\langle\cos^2\gamma\rangle^{1/2}\right] \tag{9}$$

and similarly for γ_D. In the limit of small amplitude vibrational motion, γ_A and γ_D are root-mean-square coordinates.

Several interesting trends can be seen in Table 5. γ_D is essentially the angle made by the proton donor (HF or HCl) with respect to a linear hydrogen bond axis as reference; e.g., the F ... F axis for $(HF)_2$. In most of the cases listed, γ_D is roughly 20°, which is a reasonably order-of-magnitude for the root-mean-square

Table 5. Monomer orientations from nuclear hyperfine interactions. γ_D and γ_A are the angles made by the proton donor and the proton acceptor symmetry axes, respectively, with the hydrogen bond axis.

Molecule	γ_D degrees)	γ_A (degrees)	Source[a]	Ref.
$(HF)_2$	23	70	D—eqQ	[11]
HF · HCl	22.88		Cl—eqQ	[31]
HF · DCl	19.23, 23		Cl, D—eqQ	[31]
$NH_3 · H_2O$		23	N—eqQ	[80]
OC · HF	21.90		HF—SS	[45]
OC · DF	22.17		D—eqQ	[45]
OC · HCl	23.0		Cl—eqQ	[44]
OC · DCl	20.3		Cl—eqQ	[44]
$CO_2 · HF$	25.2		HF—SS	[33]
$CO_2 · DF$	22.3, 24.4		DF—SS, D—eqQ	[33]
NCCN · HF	20.1	9, 18[b]	HF—SS, N—eqQ	[47]
HCN · HCl	21.11	18.01	Cl, N—eqQ	[54]
HCN · DCl	19.75	17.92	Cl, N—eqQ	[54]
$(HCN)_2$	11.04	17.34	N—eqQ	[27]
$N_2 · HF$	25.5	16	HF—SS, N—eqQ	[48]
$N_2O · HF$	31.1		HF—SS	[32]
$N_2O · DF$	32.42		D—eqQ	[32]
$C_2H_2 · HF$	20		HF—SS	[49]
$C_2H_2 · DF$	22		D—eqQ	[49]
$C_2H_4 · HF$	20		HF—SS	[52]
$C_2H_4 · DF$	22		D—eqQ	[52]
$C_2H_4 · HCl$	15.46, 15.57[c]		Cl—eqQ	[53]
$C_2H_4 · DCl$	13.99, 14.11[c]		Cl—eqQ	[53]
$C_2H_2 · HCl$	15.69, 15.00[c]		Cl—eqQ	[50]
$C_2H_2 · DCl$	14.21, 13.26[c]		Cl—eqQ	[50]
$C_3H_6 · HF$	19		HF—SS	[51]
$C_3H_6 · DF$	21		D—eqQ	[51]

[a] SS is an abbreviation for nuclear spin-spin interaction.
[b] Effective γ_A values for non-bonded and hydrogen bonded nitrogen, respectively.
[c] In-plane and out-of-plane γ_D values, respectively.

vibrational displacement from a linear, equilibrium hydrogen bond. This view is supported by the reduction in γ_D upon deuteration as determined by the *same* hyperfine interaction, such as the chlorine quadrupole coupling for HF · HCl and HF · DCl or the spin-spin interaction for OCO · HF and OCO · DF. Crudely, γ_D for HF or HCl might be expected to decrease by a factor of $2^{1/4}$ upon deuteration, assuming harmonic oscillator behavior. The decreases in Table 5 are actually less than this, presumably due to anharmonic effects and the fact that γ_D is not in general a normal coordinate.

Further support for this view can be found from the electric dipole moments. Although the monomer moments are strongly perturbed by formation of a hydrogen bond, *changes* in the dipole moment upon deuteration should be less sensitive to such perturbations. Table 4 shows an increase in μ_a of 0.03–0.08 D upon deuterating a hydrogen-bonded HF or HCl, although the free monomer moments decrease by only 0.008 D and 0.005 D respectively. This can be explained by adopting a harmonic oscillator model in which, for HF,

$$\mu_{DF} - \mu_{HF} \cong \mu^0 \cos \gamma_{DF} - \mu^0 \cos \gamma_{HF} \cong \mu^0 \gamma_{HF}^2 \left(\frac{1}{2} - \frac{1}{2\sqrt{2}} \right). \tag{10}$$

The ratio of γ_{HF} to γ_{DF} was assumed to be $2^{1/4}$ in deriving Eqn. (10). For $\gamma_{HF} \doteq 20$–$30°$, Eqn. (10) gives a dipole moment difference of 0.03–0.08 D, as observed. Thus, the dipole moment data and the hyperfine interactions support a nearly linear equilibrium hydrogen bond, with large amplitude vibrational displacements from equilibrium. It must be emphasized that none of the experimental evidence discussed so far, including this, precludes the hydrogen bond from having a small departure, on the order of 5°, from linearity at equilibrium.

Table 5 also shows rather clearly that the hydrogen bond is perturbing the monomer properties, in addition to the electric dipole moment. When γ_D is calculated from an HF spin-spin interaction for a dimer such as OC · HF and compared to γ_D from a deuterium quadrupole coupling measurement (OC · DF), γ_D shows a small increase in angle for the DF species relative to HF, rather than the anticipated decrease. It is expected that the HF bond length will increase slightly upon hydrogen bonding, presumably causing $\langle 1/R^3 \rangle$ to decrease and the effective angle from Eqn (6) to increase, roughly 2° for a 0.01 Å increase in R. The effective angle γ_D is then an upper bound to the actual angle. The experimental evidence in Table 5 suggests that γ_D from deuterium quadrupole coupling measurements is likewise an upper bound to the true angle and is even more perturbed, since it is typically larger than γ_D from the spin-spin interaction. Since the quadrupole coupling constant depends on the electric field gradient at the nucleus in question, the reasons for this are less apparent. Semi-empirical calculations [84] suggest that the monomer eq Q for DF decreases with increasing bond length, leading to an effect similar to the spin-spin interaction. Charge transfer and polarization effects could also cause substantial changes in the field gradient, and so the origin of this effect is not clear.

4 Vibrational Information

The scarcity of vibrational spectra represents a major gap in our understanding of weakly-bound complexes. With only very sparse infrared and photodissociation vibrational spectra for the gas phase clusters currently available [85-88], rotational spectra have been analysed to provide vibrational information. There are several sources of this information. The centrifugal distortion constants, typically D_J and D_{JK}, are functions of the vibrational force constants. Secondly, the nuclear hyperfine interactions provide accurate root-mean-square displacements which, again, depend on the vibrational force field. A third method involves measuring the intensity of vibrational satellites in the rotational spectrum and calculating the corresponding vibrational frequencies from the Boltzmann factor. A collection of force constant data from these methods is shown in Table 6.

Table 6. Force constants for hydrogen bonded molecules. Also given is the well depth parameter, ε, from Eqn. (11) and (12).

Molecule	k_δ (N/M)	$k_\gamma{}^a$ (10^{-20} J/rad.^{2})	ε (cm^{-1})	Ref.
$(HF)_2$	13.6		740	11)
$(H_2O)_2$	11.8		733	13)
$H_2S \cdot HF$	12		886	34)
$OC \cdot HF$	10.8		990	45)
$OC \cdot HCl$	4.46	1.46	569	44)
$OC \cdot HBr$	3.30		469	46)
$HCN \cdot HF^b$	16–26	6.3	2070^d	18)
$CH_3CN \cdot HF^c$	22.5	2.4		19)
$HCN \cdot HCl$	11.2		1199	54)
$HCN \cdot HBr$	8.4		1016	54)
$(HCN)_2$	11		1540	27)
$HF \cdot HCl$	5.6	0.66	455	31)
$C_3H_6 \cdot HF$	23		1871	51)
$C_3H_6 \cdot HCl$	8.7		959	51)
$C_2H_2 \cdot HCl$	6.9			50)
$C_2H_4 \cdot HCl$	6.6		627	53)
$N_2 \cdot HF$	7.07		607	48)
$CO_2 \cdot HF$	2.1	2.2		33)
$SCO \cdot HF$	2.8			33)

^{a} Hydrogen bond bending force constant.
^{b} $k_B = 3.7 \times 10^{-20}$ J/rad.^{2} for proton acceptor bend; high frequency stretching constant $k_s = 778$ N/m.
^{c} $k_B = 8.0 \times 10^{-20}$ J/rad.^{2}; $k_s = 742$ N/m.
^{d} $D_e = 2180$ cm^{-1} from microwave intensity measurements.

The low frequency ($v_\sigma \sim 100$ cm^{-1}) stretching force constants, k_σ, in Table 6 are in all cases, except $HCN \cdot HF$ and $CH_3CN \cdot HF$, calculated from the centrifugal distortion parameter, D_J, using a pseudo-diatomic model in which the monomers of the complex are treated as point masses m_1 and m_2:

$$D_J = \frac{4B^3}{v_\sigma^2} \tag{10a}$$

$$v_\sigma = \frac{1}{2\pi} \sqrt{\frac{k_\sigma}{\mu}} \tag{10b}$$

Thomas R. Dyke

$$\mu = \frac{m_1 m_2}{m_1 + m_2} \qquad (10c)$$

For HCN · HF and CH$_3$CN · HF, a more complete treatment was carried through in which the input data was the high frequency stretching mode, v_s, from the IR spectrum along with v_σ determined from microwave intensity measurements of vibrational satellites [18,19]. Two *caveats* concerning k_σ derived from these methods must be noted. First, the proton accepting molecule in many instances is not even approximately a point mass, causing one source of error in deriving k_σ from Eqn. (10a). Secondly, even if k_σ can reasonably be defined by Eqn. (10b), it will be a composite of force constants [18] and not simply the pure hydrogen bond stretching force constant.

The k_σ values in Table 6 follow a reasonable pattern with "normal" hydrogen bonded dimers such as (HF)$_2$ and HCN · HF in the range 10–20 N/m while what would normally be considered as weaker bonds — C$_2$H$_2$ · HCl, HF · HCl — generally fall below 10 N/m. Such a connection between k_σ and the hydrogen bond binding energy has been suggested [44,45,48,51] from a diatomic molecule model using a Lennard-Jones 6/12 potential:

$$V(R) = \varepsilon \left[\left(\frac{R_e}{R} \right)^{12} - 2 \left(\frac{R_e}{R} \right)^6 \right] = -\varepsilon + \frac{36E}{R_e^2} (R - R)^2 + ... \qquad (11)$$

Hence,

$$\varepsilon = \frac{R_e^2 k_\sigma}{72} \qquad (12)$$

The well depth, ε, calculated from Eqn. (12) is given for a number of molecules in Table 6. If R_e is choosen as the distance between the center-of-masses of the two monomers, the trend mentioned above is reasonably well borne out, but ε for molecules such as (HF)$_2$, (H$_2$O)$_2$ and H$_2$S · HF have values less than half that of HCN · HF or more surprisingly, cyclopropane · HF. They are also less than half that predicted by *ab initio* calculation [7].

It is particularly interesting that the cyclopropane · HF molecule does have such a large value for k_σ. This fact along with the 3.02 Å distance between the F and the center of the nearest C—C bond and the root-mean-square angular HF deviation of 19° all suggest that this molecule has a normal hydrogen bond strength, comparable to (HF)$_2$ or (H$_2$O)$_2$. A second interesting result is the very low k_σ value for OCO · HF. Although approximations in the model used may be the explanation, this result is rather surprising given the reasonably short O ... F distance of about 2.83 Å. This fact along with the linear structure, which was not expected based on the electron pair donor-acceptor arguments given earlier, underscores the anomolous nature of this particular complex.

In addition to k_σ values, force constants (Table 6) for the hydrogen bond bending have been calculated from root-mean square angles from hyperfine interactions, using a triatomic model:

$$\Delta\gamma = \frac{1}{\sqrt[4]{k_\gamma I_\gamma}} \qquad (13)$$

where k_γ is the bending force constant and I_γ the corresponding moment-of-inertia. For $HCN \cdot HF$ and $CH_3CN \cdot HF$, a more complete set of force constants [18, 19] (Table 6) were determined by using the centrifugal distortion constant D_{JK}, along with infrared spectra and microwave intensity results.

III Inert Gas — Hydrogen Halide Dimers

1 Vibrationally Averaged Structures

Complexes of an inert gas atom bound to a hydrogen halide molecule have been very actively studied in recent years by molecular beam techniques. The very weak bonds (0.1–1 kcal/mole) for these molecules distinguish them from the dimers discussed in previous sections. In one sense they are important as prototype hydrogen bonds, although the binding may turn out to be quite different than for normal hydrogen bonds. In another sense, they serve as useful models for developing techniques to handle large amplitude internal motions. The weak binding ensures that rigid rotor models will not provide structural information which is isotope independent or which is characteristic of the equilibrium geometry. The systems are simple enough that they are amenable to more rigorous treatments of the internal motions based on potential surface calculations.

Structural parameters for inert gas-hydrogen halide complexes are given in Table 7. The heavy atom separation, R, in Table 7 is derived from the effective

Table 7. Vibrationally averaged structural parameters for rare gas, hydrogen halide complexes. R is the heavy atom separation and the γ's are effective RG—X—H angles based on $\langle \cos \gamma \rangle$ and $\langle \cos^2 \gamma \rangle$.

Molecule	R (Å)	γ_{P_1} (deg.)	γ_{P_2} (deg.)	μ_a (Debye)	Ref.
Ar · HF	3.5095	48.3	41.3	1.3353	35)
Ar · DF	3.4605	34.0	33.3	1.6771	35)
^{84}Kr · HF	3.6467		38.9		55)
^{84}Kr · DF	3.6399		30.3		55)
^{129}Xe · HF	3.8152	37.9	35.6	1.7359	36)
^{129}Xe · DF	3.8111	28.1	29.4	1.9537	36)
Ne · H^{35}Cl		78.5	51.5		65)
Ne · D^{35}Cl	3.7992	65.0	47.0	0.47002	37)
Ar · H^{35}Cl	4.0065	47.5	41.5	0.8114	12)
Ar · D^{35}Cl	4.0247	33.5	33.7	1.0036	12)
^{84}Kr · H^{35}Cl	4.1117	39.8	37.7	0.95669	38,15)a
^{84}Kr · D^{35}Cl	4.1258	29.1	30.5	1.10637	38,15)a
^{129}Xe · H^{35}Cl	4.2753		34.5		57)
^{129}Xe · D^{35}Cl	4.2875		27.9		57)
Ar · H^{79}Br	4.1483	50.4	42.1	0.5752	39,43)a
Ar · D^{79}Br	4.1747	33.2	34.4	0.7563	39,43)a
^{84}Kr · H^{79}Br	4.2572		38.0		43)
^{84}Kr · D^{79}Br	4.2809		31.0		43)

a Data of Ref. 38) and 39) given here.

Thomas R. Dyke

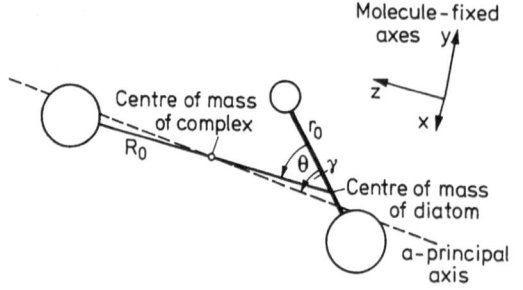

Fig. 5. Coordinate system used to describe inert gas, hydrogen halide complexes. (Reproduced from Ref. [39].)

rotational constant, $\bar{B}_0 = (B_0 + C_0)/2$, determined from the rotational spectrum. Referring to the coordinate system in Fig. 5, the distance R_0 can be determined to a first approximation by the diatomic molecule expression

$$\bar{B}_0 \cong B_d = \hbar^2/2MR_0^2 \tag{14}$$

where M is the reduced mass of the complex, treating the hydrogen halide molecule as a point mass. To a better approximation [39]

$$\bar{B}_0 \cong B_d[1 - B_d(1 + \langle\cos^2\theta\rangle)/2b_0] \tag{15}$$

where $b_0 = \hbar^2/2\mu_{HX}r_0^2$ and can be taken as the free HX monomer rotational constant. From the geometry and from $\langle\cos^2\theta\rangle$ calculated from the nuclear hyperfine interactions, Eqn. (6), $\langle\cos^2\theta\rangle$ can be evaluated for use in Eqn. (15). Further corrections are discussed in Ref. 39. Once R_0 has been determined, the (vibrationally averaged) value of the heavy atom separation is easily calculated.

In addition to R, electric dipole moment a-components are given in Table 7. Assuming that charge transfer can be neglected, this dipole moment is simply the projection of the hydrogen halide moment along the a-axis plus induced moment terms [Eqn. (8)],

$$\mu_a \cong \mu_{HX}\langle P_1(\cos\gamma)\rangle [1 + 2\alpha/R_0^3] + [3\theta_{HX}\langle P_2(\cos\gamma)\rangle/R_0^4] + \ldots \tag{16}$$

where P_1 and P_2 are Legendre polynormials, is the inert gas polarizability and θ_{HX} is the hydrogen halide quadrupole moment. $\langle\cos^2\gamma\rangle$ and hence $\langle P_2(\cos\gamma)\rangle$ is determined accurately from the nuclear hyperfine interactions, allowing $\langle P_1(\cos\gamma)\rangle$ to be approximately calculated from Eqn. (16). These expectation values are listed in Table 7 as

$$\gamma_{P_1} = \arccos[\langle\cos\gamma\rangle^{1/2}]$$
$$\gamma_{P_2} = \arccos[\langle\cos^2\gamma\rangle^{1/2}] \tag{17}$$

One ambiguity arising from this procedure is that the γ's defined by Eqn. (17) can be either acute, as drawn in Fig. 5, or obtuse, since $\cos\gamma$ is invariant to $\gamma \to 180° - \gamma$. R_0 and R are weakly changed by the acute or obtuse choice of γ. It is generally found that R decreases as the mass of the heavy atom increases for an acute choice of γ; e.g., R = 4.2753 Å for ^{129}XeH^{35}Cl and 4.2747 Å for ^{129}XeH^{37}Cl if γ is acute [57]. By analogy with more strongly bound molecules, such a decrease is expected, and the acute choice of γ has been made for all of these complexes. Similar, but even larger effects are observed upon deuterium substitution; the Xe—F distance of Xe · HF increases by 0.082 Å upon deuterium substitution for the obtuse choice of γ, but decreases by 0.004 Å for the acute choice [36]. The situation is not entirely satisfactory, since it rests primarily on analogies with isotope effects for more strongly bound molecules, in which vibrational anharmonicity is the determining factor.

The large amplitude vibrational effects are shown quite dramatically by the differences in γ based on $\langle\cos^2\gamma\rangle$ and $\langle\cos\gamma\rangle$, and also in the large difference in R and γ upon deuterium substitution. The Ne · HCl system is particularly interesting in that γ_{P_1} and γ_{P_2} are close to the free rotor limits of 90° and 54.74°, respectively. The decrease in γ upon deuteration occurs in all cases, and it is generally assumed that the equilibrium geometry is closer to a linear configuration than the vibrationally averaged structures.

The vibrational effects have been roughly modelled by using D_J and γ_{P_2} in conjunction with Eqn. (10a) and Eqn. (13), respectively, to determine hydrogen bond bending and stretching force constants, k_σ and k_b. These constants and the associated harmonic frequencies are given in Table 8. Since the results are not isotope idependent, the model has been extended in some cases to include radial-angular coupling [39, 43, 57]:

$$V(R, \theta) = \frac{1}{2} k_\sigma^0 (R - R_e)^2 + \frac{1}{2} k_b^0 \theta^2$$
$$+ k_3\theta^2(R - R_e) + k_4\theta^2(R - R_e)^2 \tag{18}$$

The observed stretching and bending force constants from Eqn. (10) and (13) are then expressed as

$$k_\sigma = k_\sigma^0 + 2k_4\langle\theta^2\rangle \tag{19}$$
$$k_b = k_b^0 + 2k_3\langle R - R_e\rangle$$

$\langle\theta^2\rangle$ is known from the angular information in Table 7 and $\langle R - R_e\rangle$ can be estimated (see Ref. 43) from the isotope shifts in R along with Eqn. (11), allowing k_σ^0 and k_b^0 to be evaluated (Table 8). k_σ^0 and k_b^0 are considerably changed from k_σ and k_b, but it is difficult to determine how meaningful they are. Presumably k_σ^0, k_b^0 and a well-depth, ε, calculated from k_σ^0 and Eqn. (12) give qualitatively useful information. The well depths increase in magnitude as both the rare gas and the halogen become heavier. It seems likely that dispersion forces along with electro-statically induced attraction are the dominant attractive forces for these complexes, rather than charge transfer, since the well-depth does not correlate with the hydrogen bonding ability of the hydrogen halide molecule.

2 Vibrational Potential Calculations

The inert gas hydrogen halide dimers have been extensively used in calculations whose goal is to fit a vibrational potential energy surface to a variety of experimental

Table 8. Vibrational frequencies and force constants for hydrogen bond stretching and bending. The Lennard-Jones 6/12 potential well depth parameter, ε, is also given.

Molecule	ω_δ (cm^{-1})	k_σ (N/m)	k_σ^0 (N/m)	ω_b (cm^{-1})	k_b (10^{-20} J/rad.^{2})	k_b^0	ε (cm^{-1})	Ref.
Ar · HF	42.5	1.42		79.8	0.31			35)
Ar · DF	45.6	1.68		65.5	0.39			35)
^{84}Kr · HF	43.7	1.82		92.1	0.42			55)
^{84}Kr · DF	46.1	2.10		74.7	0.53			55)
^{129}Xe · HF	45	2.1		100	0.5			36)
^{129}Xe · DF	47	2.4		80	0.6			36)
Ne · D^{35}Cl	21.8	0.38			0.76			37)
Ar · H^{35}Cl	32.4	1.17	1.64^a			0.15^a	184^a	12)
Ar · D^{35}Cl	34.4	1.34						12)
^{84}Kr · H^{35}Cl	32.3	1.55	2.03	52.9	0.13	0.22	240	15)
^{84}Kr · D^{35}Cl	33.6	1.71		39.8	0.14			15)
^{129}Xe · H^{35}Cl	33.9	1.90	2.33	56.8	0.31	0.32	298	57)
^{129}Xe · D^{35}Cl	34.8	2.05		44.2	0.37			57)
Ar · H^{79}Br	22.1	0.76	1.66	30.8	0.11	0.14	206	43)b
Ar · D^{79}Br	26.0	1.06		23.4	0.13			43)b
^{84}Kr · H^{79}Br	23.4	1.32	1.92	37.7	0.17	0.21	247	43)
^{84}Kr · D^{79}Br	25.0	1.52		28.7	0.19			43)

^{a} Calculated in Ref. 57).
^{b} Also see Ref. 39).

parameters, including rotational line broadening cross sections, virial coefficients and molecular beam differential cross sections in addition to the microwave spectra discussed above [60-65, 89-91]. The microwave spectra are obtained for the vibrational ground state and are therefore most sensitive to the potential surface in the vicinity of the energy minimum. A number of calculations have been carried out to fit this high precision data using a Born-Oppenheimer type separation of radial and angular coordinates. By employing this separation, calculations can be performed which are efficient and inexpensive [60, 61] compared to a close-coupling calculation [90].

In the treatment originally discussed by Homgren et al. [60, 61], the vibration-rotation wave function was factorized so that the angular part depended parametrically on R_0 (Fig. 5). The Hamiltonian could then be divided to give an angular part hat was readily expressed as a matrix and diagonalized. The resulting eigenvalues gave an effective radial potential, $U(R_0)$, analogous to the effective electronic potential curves from a conventional Born-Oppenheimer treatment. The radial Schroedinger equation was then solved by numerical means. The potential surface can be expressed as a sum over Legendre polynomials for these calculations,

$$V(R, \theta) = \sum_k V_k(R) \, P_k(\cos \theta) \tag{20}$$

although other parameterizations have been found useful [64]. These methods have been substantially extended by Hutson and Howard [63-65] who have introduced corrections for the breakdown in the Born-Oppenheimer separation, and also introduced an alternate factorization in which the radial problem is solved first.

This latter method should be particularly useful for cases where the stretching frequency becomes higher than the bending frequency.

It is outside the scope of this review to give a detailed assessment of these calculations. However, in general, they have given satisfactory fits to the microwave spectral data, and in some cases, potential surfaces have been found which fit scattering data, rotational line broadening and virial coefficient data as well [64]. As an example, the fit of a potential with 9 free parameters for Ar · HCl and 5 free parameters for Ne · HCl, Kr · HCl and Xe · HCl to molecular beam spectroscopic data is given in Table 9. In addition to an excellent fit of this data, the potentials in question also fit rotational line broadening cross sections quite accurately.

Table 9. Fits to parameters from molecular beam spectroscopy using a Born-Oppenheimer approximation. D_θ represents the centrifugal distortion of the quadrupole coupling interaction. (Reproduced from Ref. [65])

	B/MHz	D_J/kHz	$\langle P_1(\cos\theta)\rangle$	$\langle P_2(\cos\theta)\rangle$	D_θ/p.p.m.
Ne · H^{35}Cl					
Experimental	2731.44	—	0.200	0.0812	—
M5 potential	2732.9	206.9	0.188	0.0782	12.3
Uncertainty	6.0	5.0	0.008	0.004	10.0
Ne · D^{35}Cl					
Experimental	2696.44	183.07	0.4203	0.1972	73.2
M5 potential	2695.7	183.3	0.425	0.1968	53.4
Uncertainty	6.0	5.0	0.008	0.004	10.0
Ar · H^{35}Cl					
Experimental	1678.511	20.0	0.6733	0.3406	23.7
M5 potential	1679.5	20.2	0.653	0.340	21.0
Uncertainty	2.0	0.5	0.02	0.002	2.0
Ar · D^{35}Cl					
Experimental	1656.19	17.1	0.8331	0.5370	19.0
M5 potential	1655.5	16.7	0.808	0.538	19.1
Uncertainty	2.0	0.5	0.027	0.002	3.0
Kr · H^{35}Cl					
Experimental	1200.624	7.34	0.7659	0.4324	10.8
M5 potential	1199.7	7.66	0.731	0.429	8.8
Uncertainty	3.0	0.2	0.032	0.040	1.5
Kr · D^{35}Cl					
Experimental	1182.72	6.51	0.8799	0.6055	10.7
M5 potential	1182.3	6.37	0.844	0.607	7.3
Uncertainty	3.0	0.2	0.040	0.002	3.0
Xe · H^{35}Cl					
Experimental	994.145	3.813	0.8297	0.5141	—
M5 potential	994.2	3.94	0.792	0.513	5.5
Uncertainty	1.5	0.1	0.05	0.002	2.0
Xe · D^{35}Cl					
Experimental	978.48	3.452	—	0.6650	—
M5 potential	977.6	3.31	0.872	0.666	4.6
Uncertainty	1.5	0.1	0.05	0.002	2.0

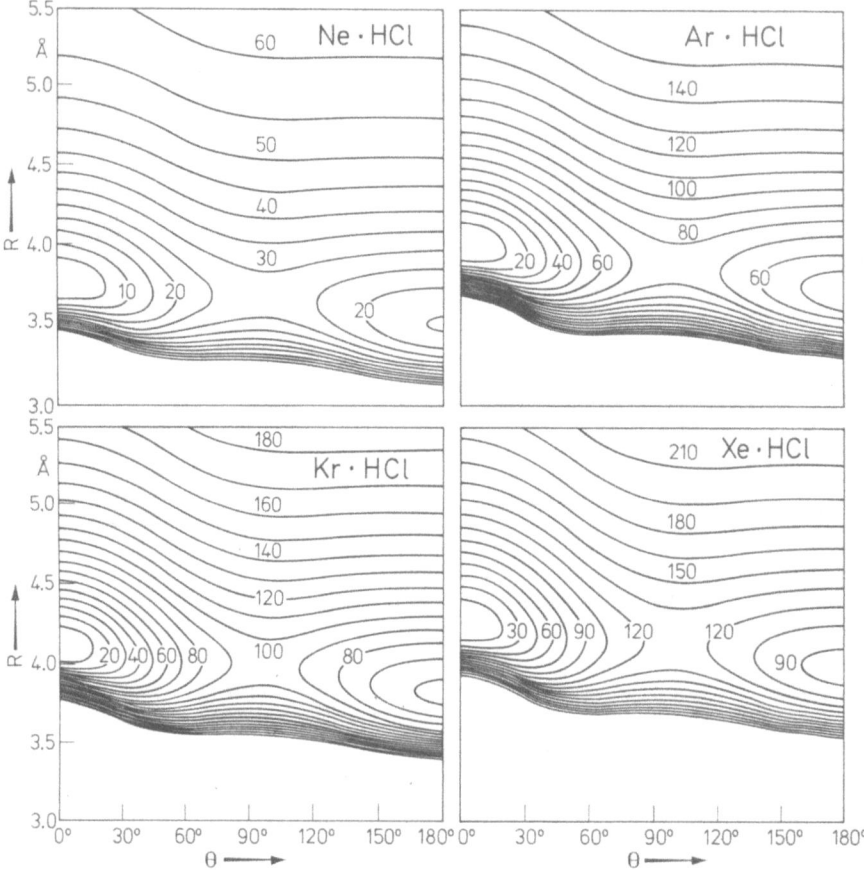

Fig. 6. Contour plots (cm^{-1}) of potential surfaces in Table 9. (Reproduced from Ref. [65])

Contour plots for these potentials are shown in Fig. 6. It is particularly interesting that the minimum occurs at a linear RG · HX configuration, which is generally the case, but there is also a secondary minimum at a linear RG · XH configuration.

E Discussion

It is now possible to determine precise rotational spectra for hydrogen bonded molecules of moderate size and with even very small stabilization energies. Rotational constants, centrifugal distortion constants, electric dipole moments and nuclear hyperfine interactions have been measured for a considerable number of dimers using various microwave and molecular beam techniques.

The picture of the hydrogen bond that has emerged features hydrogen bonds that are linear at equilibrium within error limits of about 10°. One of the interesting questions for future work will be to refine that picture to see if any small non-linearities exist. The hydrogen bonds are generally longer for gas phase dimers than for condensed phases. This effect can be attributed to cooperative effects; i.e., upon formation of a hydrogen bond, charge transfer occurs which strengthens the next hydrogen bond to form. Such effects must be included in models of condensed phases. In this regard, microwave spectra for trimers and higher polymers will be quite important. Although electric deflection [92-94] and photodissociation experiments [87, 88] indicate that hydrogen-bonded trimers and higher polymers form cyclic rings, precision structural information is needed to decide this question.

For normal hydrogen bonded dimers, an electron pair donor-acceptor model which exploits traditional chemical concepts of lone-pair directionality has been successful in rationalizing the orientation of the electron donor constituent and the linear hydrogen bonds. The similarities in bond angles between the first and second row hydrides and their corresponding dimers are quite striking in support of these ideas. In the case of inert gas, hydrogen halide complexes, the very weak bonds are probably due to dispersion and electrostatically induced attractions rather than charge transfer.

A certain amount of information concerning the vibrational potential has been extracted from rotational spectra. Developing molecular beam methods for systematically acquiring vibrational spectra for the low frequency intermolecular modes of dimers is an important problem to be attacked. Such information along with the developing techniques for theoretically handling the internal motions of "floppy" molecules will lead to an accurate picture of the vibrational potential energy surface and equilibrium properties of hydrogen bonded molecules.

F Acknowledgements

The author would like to thank B. J. Howard, W. Klemperer, J. S. Muenter and S. E. Novick for substantial help and information concerning this area of research.

G References

1. G. C. Pimentel, A. D. McClellan: *The Hydrogen Bond*, Freeman, San Francisco, 1960
2. F. Franks: *Water, a Comprehensive Treatise*, Plenum, New York, 1972, V. 1–6
3. M. D. Joesten, L. J. Schaad: *Hydrogen Bonding*, Marcel Dekker, New York, 1974
4. Ref. 2, V. 4–6
5. H. R. Carlon, C. A. Harden: Appl. Optics *19*, 1776 (1980)
6. D. T. Lewellyn Jones, R. J. Knight, H. A. Gebbie: Nature *274*, 876 (1978)
7. J. D. Dill, L. C. Allen, W. C. Topp, J. A. Pople: J. Am. Chem. Soc. *97*, 7220 (1975)
8. K. Morokuma: Accts. Chem. Res. *10*, 295 (1977)
9. O. Matsuoka, E. Clementi, M. Yoshimine: J. Chem. Phys. *64*, 1351 (1976)
10. C. C. Costain, G. P. Srivastava: J. Chem. Phys. *35*, 1903 (1961); *41*, 1620 (1964)
11. T. R. Dyke, B. J. Howard, W. Klemperer: J. Chem. Phys. *56*, 2442 (1972)
12. S. E. Novick, P. Davies, S. J. Harris, W. Klemperer, J.: Chem. Phys. *59*, 2273 (1973)
13. a) T. R. Dyke: J. Chem. Phys. *66*, 492 (1977);
    b) T. R. Dyke, K. M. Mack, J. S. Muenter: J. Chem. Phys. *66*, 498 (1977);
    c) J. A. Odutola, T. R. Dyke: J. Chem. Phys. *72*, 5062 (1980)
14. D. J. Millen: J. Mol. Struct. *45*, 1 (1978)
15. T. J. Balle, E. J. Campbell, M. R. Keenan, W. H. Flygare: J. Chem. Phys. *72*, 922 (1980)

16. E. M. Bellott, Jr., E. B. Wilson: Tetrahedron *31*, 2896 (1975)
17. A. C. Legon, D. J. Millen, S. C. Rogers: J. Molec. Spectrosc. *70*, 209 (1978)
18. A. C. Legon, D. J. Millen, S. C. Rogers: Proc. R. Soc. Lond. *A370*, 213 (1980)
19. J. W. Bevan, A. C. Legon, D. J. Millen, S. C. Rogers: Proc. R. Soc. Lond. *A370*, 239 (1980)
20. A. S. Georgiou, A. C. Legon, D. J. Millen: Proc. R. Soc. Lond. *A370*, 257 (1980)
21. A. C. Legon, D. J. Millen, S. C. Rogers: J. Mol. Struct. *67*, 29 (1980)
22. J. W. Bevan, Z. Kisiel, A. C. Legon, D. J. Millen, S. C. Rogers: Proc. R. Soc. Lond. *A372*, 441 (1980)
23. A. S. Georgiou, A. C. Legon, D. J. Millen: J. Mol. Struct. *69*, 69 (1980)
24. A. S. Georgiou, A. C. Legon, D. J. Millen: Proc. R. Soc. Lond. *A372*, 511 (1981)
25. A. C. Legon, D. J. Millen, P. J. Mjoberg: Chem. Phys. Letters *47*, 589 (1977)
26. C. H. Townes, A. W. Schawlow: *Microwave Spectroscopy*, McGraw-Hill, New York, 1955
27. L. W. Buxton, E. J. Campbell, W. H. Flygare: Chem. Phys. *56*, 399 (1981)
28. J. B. Anderson, in: *Molecular Beams and Low Density Gas Dynamics* (P. P. Wegener, ed.) Dekker, New York, 1974, Vol. 4, Chap. 1
29. T. R. Dyke, J. S. Muenter: Properties of Molecules from Molecular Beam Spectroscopy in *Internat. Rev. Sci.*, Phys. Chem. Ser. 2, Vol. 2 (A. D. Buckingham, ed.), Butterworths, London, 1975
30. T. C. English, J. C. Jorn in: Methods of Experimental Physics (D. Williams, ed.), Vol. 3, Academic Press, New York, 1973
31. K. C. Janda, J. M. Steed, S. E. Novick, W. Klemperer: J. Chem. Phys. *67*, 5162 (1977)
32. C. H. Joyner, T. A. Dixon, F. A. Baiocchi, W. Klemperer: ibid. *74*, 6550 (1981)
33. F. A. Baiocchi, T. A. Dixon, C. H. Joyner, W. Klemperer: ibid. *74*, 6544 (1981)
34. R. Viswanathan, T. R. Dyke: ibid. *77*, 1166 (1982)
35. T. A. Dixon, C. H. Joyner, F. A. Baiocchi, W. Klemperer: J. Chem. Phys. *74*, 6539 (1981).
36. F. A. Baiocchi, T. A. Dixon, C. H. Joyner, W. Klemperer: ibid. *75*, 2041 (1981)
37. A. E. Barton, D. J. B. Howlett, B. J. Howard: Molec. Phys. *41*, 619 (1980)
38. A. E. Barton, T. J. Henderson, P. R. R. Langridge-Smith, B. J. Howard: Chem. Phys. *45*, 429 (1980)
39. K. C. Jackson, P. R. R. Langridge-Smith, B. J. Howard: Molec. Phys. *39*, 817 (1980)
40. E. J. Campbell, L. W. Buxton, T. J. Balle, W. H. Flygare: J. Chem. Phys. *74*, 813 (1981)
41. E. J. Campbell, L. W. Buxton, T. J. Balle, M. R. Keenan, W. H. Flygare: J. Chem. Phys. *74*, 829 (1981)
42. T. J. Balle, W. H. Flygare: Rev. Sci. Instrum. *52*, 33 (1981)
43. M. R Keenan, E. J. Campbell, T. J. Balle, L. W. Buxton, T. K. Minton, P. D. Soper, W. H. Flygare: J. Chem. Phys. *72*, 3070 (1980)
44. P. D. Soper, A. C. Legon, W. H. Flygare: ibid. *74*, 2138 (1981)
45. A. C. Legon, P. D. Soper, W. H. Flygare: ibid. *74*, 4944 (1981)
46. M. R. Keenan, T. K. Minton, A. C. Legon, T. J. Balle, W. H. Flygare: Proc. Natl. Avad. Sci. USA *77*, 5583 (1980)
47. A. C. Legon, P. D. Soper, W. H. Flygare: J. Chem. Phys. *74*, 4936 (1981)
48. P. D. Soper, A. C. Legon, W. G. Read, W. H. Flygare: ibid. *76*, 292 (1982)
49. W. G. Read, W. H. Flygare: ibid. *76*, 2238 (1982)
50. A. C. Legon, P. D. Adrich, W. H. Flygare: ibid. *75*, 625 (1981)
51. L. W. Buxton, P. D. Aldrich, J. A. Shea, A. C. Legon, W. H. Flygare: J. Chem. *75*, 2681 (1981)
52. J. A. Shea, W. H. Flygare: ibid. *76*, 4857 (1982)
53. P. D. Aldrich, A. C. Legon, W. H. Flygare: ibid. *75*, 2126 (1981)
54. A. C. Legon, E. J. Campbell, W. H. Flygare: ibid. *76*, 2267 (1982)
55. L. W. Buxton, E. J. Campbell, M. R. Keenan, T. H. Balle, W. H. Flygare: Chem. Phys. *54*, 173 (1981)
56. E. J. Campbell, M. R. Keenan, L. W. Buxton, T. J. Balle, P. D. Soper, A. C. Legon, W. H. Flygare: Chem. Phys. Letters *70*, 420 (1980)
57. M. R. Keenan, L. W. Buxton, E. J. Campbell, T. J. Balle, W. H. Flygare: J. Chem. Phys. *73*, 3523 (1980)
58. H. C. Longuet-Higgins: Molec. Phys. *6*, 445 (1963)
59. J. A. Odutola, D. Prinslow, T. R. Dyke: To be published

60. S. E. Novick, K. C. Janda, S. L. Holmgren, M. Waldman, W. Klemperer: J. Chem. Phys. *65*, 1114 (1976)
61. S. L. Holmgren, M. Waldman, W. Klemperer: J. Chem. Phys. *69*, 1661 (1978); *67*, 4414 (1977)
62. J. M. Hutson, A. E. Barton, P. R. R. Langridge-Smith, B. J. Howard: Chem. Phys. Letters *73*, 218 (1980)
63. J. M. Hutson, B. J. Howard: Molec. Phys. *41*, 1123 (1980)
64. J. M. Hutson, B. J. Howard: ibid. *43*, 493 (1981)
65. J. M. Hutson, B. J. Howard: ibid. *45*, 769, 791 (1982)
66. N. F. Ramsey: *Molecular Beams*, Oxford University Press, London, 1955
67. P. Thaddeus, L. Krisher, J. Loubser: J. Chem. Phys. *40*, 257 (1964)
68. J. Karle, L. O. Brockway: J. Am. Chem. Soc. *66*, 574 (1944)
69. J. A. Odutola, T. R. Dyke: J. Chem. Phys. *68*, 5663 (1978)
70. F. Holtzberg, B. Post, I. Fankuchen: Acta Cryst. *6*, 127 (1953)
71. R. E. Jones, D. H. Templeton: ibid. *11*, 484 (1958)
72. M. Atoji, W. N. Lipscomb: ibid. *7*, 173 (1954)
73. W. J. Siemons, D. H. Templeton: ibid. *7*, 194 (1954)
74. H. S. Frank: Proc. Royal Soc. Lond. *A247*, 481 (1958)
75. H. S. Frank and W. Y. Wen: Disc. Faraday Soc. *24*, 133 (1957)
76. A. C. Legon, P. D. Soper, M. R. Keenan, T. K. Minton, T. J. Balle, W. H. Flygare: J. Chem. Phys. *73*, 583 (1980)
77. A. R. Ubbelohde, K. J. Gallagher: Acta. Cryst. *8*, 71 (1955)
78. As quoted in ref. 45
79. F. A. Baiocchi, W. Klemperer: private communication
80. P. Herbine, T. R. Dyke: To be published
81. S. E. Novick, K. C. Janda, W. Klemperer: J. Chem. Phys. *65*, 5115 (1976)
82. A. D. Buckingham: Quart. Rev. (Lond.) *13*, 183 (1959)
83. L. C. Allen: J. Am. Chem. Soc. *97*, 6921 (1975)
84. M. Barfield, H. P. W. Gottlieb, D. M. Doddrell: J. Chem. Phys. *69*, 4504 (1978). Also see ref. 52
85. R. K. Thomas: Proc. Royal Soc. Lond. *A325*, 133 (1971)
86. R. K. Thomas: ibid. *A344*, 579 (1975)
87. M. F. Vernon, D. J. Krajnovioh, H. S. Kwok, J. M. Lisy, Y. R. Shen, Y. T. Lee: J. Chem. Phys. *77*, 47 (1982)
88. J. M. Lisy, A. Tramer, M. F. Vernon, Y. T. Lee: J. Chem. Phys. *75*, 4733 (1981)
89. W. B. Neilsen, R. G. Gordon: J. Chem. Phys. *58*, 4149 (1973)
90. A. M. Dunker, R. G. Gordon: J. Chem. Phys. *64*, 354; 4984 (1976)
91. J. M. Farrar, Y. T. Lee: Chem. Phys. Lett. *26*, 428 (1974)
92. J. A. Odutola, R. Viswanathan, T. R. Dyke: J. Am. Chem. Soc. *101*, 4787 (1979)
93. S. E. Novick, P. Davies, T. R. Dyke, W. Klemperer: J. Am. Chem. Soc. *95*, 8547 (1973)
94. T. R. Dyke, J. S. Muenter: J. Chem. Phys. *57*, 5011 (1972)

Author Index Volumes 101–120

Contents of Vols. 50–100 see Vol. 100
Author and Subject Index Vols. 26–50 see Vol. 50

The volume numbers are printed in italics

116

A.F.Williams

A Theoretical Approach to Inorganic Chemistry

1979. 144 figures, 17 tables. XII, 316 pages
ISBN 3-540-09073-8

Contents: Quantum Mechanics and Atomic Theory. – Simple Molecular Orbital Theory. – Structural Applications of Molecular Orbital Theory. – Electronic Spectra and Magnetic Properties of Inorganic Compounds. – Alternative Methods and Concepts. – Mechanism and Reactivity. – Descriptive Chemistry. – Physical and Spectroscopic Methods. – Appendices. – Subject Index.

This book is intended to outline the application of simple quantum mechanics to the study of inorganic chemistry, and to show its potential for systematizing and understanding the structure, physical properties, and reactivities of inorganic compounds. The considerable development of inorganic chemistry in recent years necessitates the establishment of a theoretical framework if the student is to acquire a sound knowledge of the subject. An effort has been made to cover a wide range of subjects, and to encourage the reader to think of further extensions of the theories discussed. The importance of the critical application of theory is emphasized, and, although the book is concerned chiefly with molecular orbital theory, other approaches are discussed. The book is intended for students in the latter half of their undergraduate studies.
(235 references)

H.J.Fischbeck, K.H.Fischbeck

Formulas, Facts and Constants

for Students and Professionals in Engineering, Chemistry and Physics
1982. XII, 251 pages
ISBN 3-540-11315-0

Contents: Basic Mathematical Facts and Figures. – Units, Conversion, Factors and Constants. – Spectroscopy and Atomic Structure. – Basic Wave Mechanics. – Facts, Figures and Data Useful in the Laboratory.

This book provides a handy and convenient source of formulas, conversion factors and constants for students and professionals in engineering, chemistry, mathematics and physics. Section 1 covers the fundamental tools of mathematics needed in all areas of the physical sciences. Section 2 summarizes the SI system (International System of Units of measurement), lists conversion factors and gives precise values of fundamental constants. Sections 3 and 4 review the basic terms of spectroscopy, atomic structure and wave mechanics. These sections serve as a guide to the interpretation of modern literature. Section 5 is a resource for work in the laboratory, listing data and formulas needed in connection with frequently used equipment such as vacuum systems and electronic devices. Material constants and other data are listed for information and as an aid for estimates or problem solving.
Formulas and tables are accompanied by examples in all those cases where their use might not be self-explanatory.

Springer-Verlag
Berlin
Heidelberg
New York
Tokyo

Lecture Notes in Chemistry

Editors: G. Berthier, M. J. S. Dewar, H. Fischer,
K. Fukui, G. G. Hall, H. Hartmann, H. H. Jaffé,
J. Jortner, W. Kutzelnigg, K. Ruedenberg, E. Scrocco

Springer-Verlag
Berlin
Heidelberg
New York
Tokyo